Withdrawn

SILENCE OF THE SPHERES

SILENCE OF THE SPHERES

The Deaf Experience in the History of Science

Harry G. Lang

BERGIN & GARVEY
Westport, Connecticut • London

Library of Congress Cataloging-in-Publication Data

Lang, Harry G.
Silence of the spheres : the deaf experience in the history of science / Harry G. Lang.
p. cm.
Includes bibliographical references and index.
ISBN 0-89789-368-9 (alk. paper)
1. Science–Social aspects–History. 2. Deaf scientists–History.
3. Deaf–Education–Science–History. I. Title.
Q175.5.L34 1994
306.4'5'0872–dc20 93-20838

British Library Cataloguing in Publication Data is available.

Library of Congress Catalog Card Number: 93-20838
ISBN: 0-89789-368-9

First published in 1994

Bergin & Garvey, 88 Post Road West, Westport, CT 06881
An imprint of Greenwood Publishing Group, Inc.

Printed in the United States of America

The paper used in this book complies with the Permanent Paper Standard issued by the National Information Standards Organization (Z39.48-1984).

10 9 8 7 6 5 4 3 2

Copyright Acknowledgments

The author and publisher gratefully acknowledge permission for use of the following material:

Photograph of Stephen Hawking and Harry G. Lang by Dr. Martha Redden, courtesy of the American Association for the Advancement of Science.

Photograph of the French Academy of Sciences from E. Maindron, *L'Académie des Sciences* (1888). Paris: Ancienne Librairie Germer Ballière.

Photograph of Charles Bonnet courtesy of Librairie Philosophique J. Vrin, from R. Savioz, *Memoires Autobiographiques de Charles Bonnet de Geneve* (1948).

Photograph of John Goodricke courtesy of the Royal Astronomical Society, from P. Moore, *Men of Stars* (1986).

Photograph of Anders Gustaf Ekeberg courtesy of the *Journal of Chemical Education*, from Mary E. Weeks, *Discovery of the Elements* (1956).

Photograph of Leo Lesquereux from E. Orton, "Leo Lesquereux," *American Geologist* 5 (May 1890).

Photograph of Fielding Bradford Meek courtesy of Yale University Press, from G. P. Merrill, *The First One Hundred Years of American Geology* (Yale University Press, 1924). Copyright © 1924 by Yale University Press.

Photograph of Thomas Meehan from *Country Life in America* (February 1902).

Photograph of Gallaudet Science Laboratory courtesy of Gallaudet University Archives.

Photograph of James H. Logan courtesy of Gallaudet University Archives, from J. E. Gallaher, *Representative Deaf Persons of the United States* (1898).

Photograph of George T. Dougherty courtesy of Gallaudet University Archives.

Photograph of Gerald M. McCarthy courtesy of the Alabama School for the Deaf and Blind, from the *Alabama Messenger* (1896).

Photograph of Annie Jump Cannon courtesy of Harvard College Observatory.

Photograph of Henrietta Swan Leavitt courtesy of Harvard College Observatory, from *Popular Astronomy* (April 1922).

Photograph of Oliver Heaviside courtesy of The Institution of Electrical Engineers, from *Heaviside Centenary Volume* (1950).

Photograph of Harold J. Conn courtesy of Jean C. Cochrane.

Photograph of Olaf Hassel courtesy of Reidun Guldal.

Photograph of Comet Jurlof-Achmarof-Hassel courtesy of Reidun Guldal.

Photograph of Wyn Owston courtesy of Dr. Wyn Owston.

Photograph of Tilly Edinger courtesy of the Museum of Comparative Zoology, Harvard University. Copyright © President and Fellows of Harvard College.

Harold J. Conn, "A Religious Scientist at the Turn of the Century: Herbert William Conn of Wesleyan University," unpublished manuscript. Courtesy of ASM Archives, Center for the History of Microbiology, University of Maryland, Baltimore County.

"The Heavenly motions are nothing but a continuous song for several voices (perceived by the intellect, not by the ear)."

Johannes Kepler
***Harmonice mundi*, 1619**

Contents

Illustrations

Preface

On a warm summer night I lie upon a grassy hill, looking up and thinking about the stillness that surrounds me. A breeze blows as I stare into the star-glittered sky. The moon is bright and I can see its larger craters. East of Mare Crisium is Crater Cannon, named in honor of a woman astronomer whose deafness, her colleagues said, increased her powers of concentration. Reflecting on my search for the "deaf experience" in the history of science, I especially think of the far side of the moon, never visible to us; some of its craters too were named for scientists and mathematicians who, in a stillness of their own, once contemplated the universe.[1]

In the constellation Perseus there shines Algol, the "demon star," whose mysterious changes in brightness were first investigated by a deaf man named John Goodricke in York, England, in the 1780s. Coincidentally, 150 years later another deaf astronomer, Olaf Hassel of Norway, was studying Perseus when he discovered his comet. Variable stars. Comets. Novae. More than six thousand of these wondrous and mysterious celestial objects have been discovered by people who lived in a world predominantly silent. Through their eyes, deaf people have indeed heard the harmonized motion of the planets, the "music of the spheres" which the Greek philosopher Pythagoras once believed the planets produced like the strings of a lyre. Edward B. Nitchie wrote, in a 1916 essay entitled "A Deaf Man's Music," "Of all the soundless harmonies of nature, the 'music of the spheres,' the wonders of the heavens by day, and especially the wonders of the heavens by night, are the mightiest and most thrilling music a deaf man's soul can hear."

I look one last time tonight into the vast expanse of space. Far out there is Beta Lyrae—two worlds orbiting one another. To me there is something wonderful about binary stars. Two worlds which share energy. Yet, usually one star in the pair is larger and brighter and it overwhelms the retina of human vision. To the unaided eye, these two stars appear as one. In the darkness, as the breeze blows softly, I think about the worlds of hearing and deaf persons. Different worlds. Companion worlds. To the unaided student of the history of science, these worlds, too, appear as one.

There has been a long silence among men and women congenitally or adventitiously deaf who have striven to have their voices heard. It has been a silence marked by prejudice and discrimination not unlike that experienced by other marginalized groups; a silence defined in part by their deafness, whose invisibility, for whatever reason, has failed to open the eyes of society to their great contributing potential; a silence perpetuated by the very fact that deaf scientists are too few in number to organize an effective cadre to systematically address their needs as scientists and as members of society.

This silence, in the spheres of human scientific endeavor, is my metaphor.

Acknowledgments

I am especially indebted to my colleagues and friends Frank Caccamise, Susan Foster, Marcia Scherer, and Bill Welsh for their repeated critiques of the manuscript and continuous support, and to T. Alan Hurwitz, Betty Morrow, and Michael Stinson for their critiques and suggestions. I also appreciate the support of my chairperson, Barbara G. McKee, in the Department of Educational Research and Development at the National Technical Institute for the Deaf.

My pursuit of this subject would never have come to fruition had it not been for the encouragement of several other colleagues. I am grateful to a friend I have never met, Francis C. Higgins, a deaf chemistry professor, now retired from Gallaudet University, who corresponded with me, sharing his own notes on the subject, and providing enthusiastic assistance whenever requested. Varadaraja V. Raman from the Rochester Institute of Technology Physics Department, an inspired writer with a love for the history of science, motivated me to delve into the literature and search for the "roots" of deaf people in scientific history.

While it is indeed a pleasant task to express gratitude, it is also a difficult one. Because so many people from libraries, archives, and scientific societies in the United States and Europe assisted me in gathering information, I hope they will understand that some receive mention only in the text and in the endnotes. Several people, however, deserve special appreciation here. Gallaudet University president I. King Jordan encouraged me to apply to the Laurent Clerc Cultural Fund, and my good friends Robert F. Panara and James DeCaro provided written support for the grant, noting emphatically the potential of this research to add a new dimension to "Deaf Studies." I thank the Gallaudet University Laurent Clerc Cultural Fund, particularly Mary Ann Pugin and Rosalyn Gannon, for the stipend which helped defray the costs of materials and assistance in translations of documents written in Swedish, Norwegian, German, and French. Thanks also to John Albertini, Rob Baker, Ebba McArt, Dominique Lepoutre, and Jean-Guy Naud for their time and energy in helping me make those translations; to Mark Benjamin, whose photographic talents brought pictures from obscure documents to life again in his prints and slides used in the book and related presentations; to

Michael Olsen, Carolyn Jones, Donna Wells, Corrine Hilton, and Marguerite Glass-Englehart at the Merrill Learning Center at Gallaudet University; and to Beth Goodrich, Virginia Stern, Martha Redden, and Michele Aldrich at the American Association for the Advancement of Science. Michele's many letters, in which she reported on her discoveries of deaf people in science, have been especially helpful to me. Kathleen Dorman at the Smithsonian Institution also was very kind in her support, as were Janice F. Goldblum and Deborah K. Williams at the National Academy of Sciences, Martha L. Hazen at the Harvard College Observatory, Lynne Hanner at the Alabama School for the Deaf and Blind, and Dorothy Sponder and Robbie McGaughan at the American Association for the Advancement of Women. I owe thanks to Jack R. Gannon for placing my query in the Gallaudet University Alumni Newsletter and to those who responded, particularly Kenneth S. Rothschild, who brought to my attention one of the two deaf Nobel laureates in science, Charles Nicolle; to the History of Science Society for publishing my query in its newsletter; and to Margaret S. Rossiter, Patricia Gossel, and Michael J. Crowe for their subsequent suggestions. These and many other people have demonstrated to me both a love for history and a collaborative spirit.

I am indebted to a number of living scientists, and to relatives of deceased scientists, who provided me with information for the present book and helped me to develop detailed biographies which will appear in my forthcoming book, *Deaf Persons in the Arts and Sciences: A Biographical Dictionary*, to be published by Greenwood Press.

I would also like to thank Richard Squires and Linda Coppola from the Wallace Memorial Library at the Rochester Institute of Technology, who answered many questions and saved me hours of labor through electronic mail and other reference services. Z. Carol Davies, Gayle Meegan, and Kim Singer were similarly helpful with requests for information and copies. Gail Kovalik, a friend and talented librarian at the National Technical Institute for the Deaf Staff Resource Center, provided a disproportionate amount of assistance in gathering materials for me. I owe a debt of gratitude to her for both her technical skills and her support and friendship.

The excellent assistance and enthusiastic support of Lynn Flint, Cathryn Lee, Karen E. Davis, and Bayard Van Hecke at the Greenwood Publishing Group made the final work on this manuscript a pleasure.

To my parents, Harry and Harriet Lang, thank you for your support through the years.

My greatest appreciation is to my wife, Professor Bonnie Meath-Lang, who shared with me many exciting moments during the research for this book, and supported me with encouragement and scholarly expertise. I dedicate this work and the two-world metaphor to her, for she has certainly convinced me that "silence" is a universal human experience.

Terms and Abbreviations

For the reader unfamiliar with deafness and deaf people in historical writing, it should be explained that the archaic terms "deaf and dumb" and "deaf-mute" are occasionally used in this book in their original contexts and should be avoided in contemporary communication. Many people who function as "hard-of-hearing" share life experiences similar to those described in this book and may be included. Following the convention described by Padden and Humphries (1988) and used by many early and contemporary writers, I will employ the general term "deaf" to refer to the audiological condition, and I will capitalize "Deaf" when referring to people who share a sign language (usually American Sign Language) and a culture. "Congenitally deaf" refers to people born deaf, whereas "adventitiously" deaf people are those who become deaf after speech and spoken language development have begun.

The "National Deaf-Mute College" became "Gallaudet College" in 1894 and was renamed "Gallaudet University" in 1986. I will use "Gallaudet College" with a few exceptions, since this was the name of the institution during the period to which most of the discussion in this book refers.

For ease of reading, I will use the following abbreviations throughout this book:

AAD — *American Annals of the Deaf* (*American Annals of the Deaf and Dumb* before 1886)
B&B — *The Buff and Blue* (Gallaudet University's alumni publication)
DA — *The Deaf American*
SW — *The Silent Worker*
VR — *The Volta Review* (*The Association Review* before 1910)

Introduction

The impetus for writing *Silence of the Spheres* was an encounter with Dr. Stephen Hawking in a small conference room in Boston in 1984. Dr. Hawking, a theoretical physicist from Cambridge University, had been invited by the American Association for the Advancement of Science to give a presentation on the subject of disabilities and the pursuit of science as a career and, as a former president of the Science Association for Persons with Disabilities, I was honored to attend his session. Dr. Hawking is one of the greatest scientific minds of this century. He has dramatically changed the field of astronomy with his ideas about black holes, singularities, and the birth of the universe. Having contracted amyotrophic lateral sclerosis, a progressive motor neuron disease more commonly known as "Lou Gehrig's disease," he was, when I met him, almost completely paralyzed. His speech was not audible to most people, and a young man who worked with him strained to hear his labored words and repeat the message for the benefit of those who had attended his lecture.

At the time I met Dr. Hawking, I had been teaching general physics for fifteen years to deaf college freshmen. With my sign language interpreter, we made an interesting foursome. When Professor Hawking spoke to me, his assistant repeated his words. My interpreter then translated this message to American Sign Language for my benefit. At one point, Dr. Hawking, a man in a wheelchair who has faced the prospect of an early death as a consequence of his disease, glanced up at me and said, "It must be difficult to be deaf."

I was shocked to see those words. As I looked away from my interpreter, my eyes met Hawking's. I did not know what to say to him. He had caught me by surprise, and I observed through nonverbal communication a sensitivity as powerful as his intellect. I awkwardly smiled back.

Later, I began to think about what Stephen Hawking had said. Over the years I had found profound deafness to be at times a minor annoyance. Although it had provided me with continual communicative challenges, I hardly felt my life was "difficult," particularly in comparison to conditions such as Dr. Hawking's. I am sure, however, that my own experiences had led me to this

Stephen Hawking and Author Harry G. Lang

attitude about deafness as much as Dr. Hawking's fortitude had helped him. I then began to think about ways of investigating the life stories of deaf men and women scientists who had encountered negative attitudes, their own and those of others, as well as the potentially isolating conditions of deafness, and who had overcome these barriers to contribute significantly to various fields of science. The use of stories about deaf scientists as a motivator for deaf students was intriguing, as well as the idea of adding a new dimension to "Deaf Studies" courses. A fascinating and extensive literature in Deaf Studies is available on writers, poets, painters, and sculptors, but no such summaries have been published on the accomplishments of deaf people in science, engineering, invention, and medicine. Deaf scientists, compelled to blend in while often feeling excluded, have nevertheless shown great initiative in these areas. Recognition has not been given adequately to them, however, because of the obscurity and invisibility of deafness. One cannot go down the list of research publications and identify the productivity of deaf scientists, or even their attendance at universities. Nor have the subtle and not-so-subtle attitude barriers faced by them been documented as well as they have been for women, African-American, and other marginalized scientists. Limited largely to industrial and trade positions, deaf people struggled to rise in science and technology through isolated experiences.

In discussing the value of biography in scientific history, Elliott (1990) has provided an interesting description of three levels of biographical writing. The first level is that of full and definitive biographies of such notables as Benjamin Franklin or Josiah Willard Gibbs with emphasis on "matters of the mind." The second level includes studies of the lives of scientists of "secondary rank" closely tied to the building of scientific institutions. The biographies of these scientists, as Elliott explained, are often hard to distinguish from the institutional histories. The third level of biographies includes the many individuals who made scientific concerns the focal point of their lives. Elliott referred to these as the "base of the pyramid" which gave support to the upper levels. In discussing this base, he also mentioned how the interface between science and society may be examined. The biographical study of the African-American scientist E. E. Just, for example, might suggest the relations between Just's abilities in science and the prejudice he faced. Hence, we might better understand the social and institutional factors, as well as intellectual ones, through such biographical work. Stephen Hawking's comment started me on an adventure in time travel—a search for the "deaf experience" in the history of science, first at the base of the pyramid, then onto other levels of biography. This search went, as Alice in Wonderland would say, "straight on like a tunnel for some way"—through archives in both the United States and Europe, those "dark wells" filled with bookshelves. In the literal sense I could also say, much more meaningfully than Alice did, that in the wonderland beyond those tunnels "there was no one to hear me." I found among the stacks of books and materials a particular name that seems to come right out of Lewis Carroll's book. The "Little Paper Family" is a veritable storehouse of information on deafness and deaf people, their experiences, frustrations, and achievements. It is a fascinating collection of magazines and newspapers with titles like *The Silent Hoosier*, *The Iowa Hawkeye*, and *The Western Pennsylvanian*, compiled by teachers, editors, and administrators of schools serving deaf students.[1] By reading the school papers, especially the more popular *The Silent Worker* (later, *The Deaf American*), and other works of the Deaf community with such archaic titles as *The Deaf Mutes' Journal* and *The Deaf Mute Howls*, one develops a feeling of community and culture among many people who are deaf, a culture marked by a constant struggle for a political, social, and intellectual "voice." Throughout this book I refer to the "literature of the Deaf community" to mean these collective works, including the "Little Paper Family."

It is not surprising that other marginalized groups, including women and economically disadvantaged people, have been described metaphorically as "invisible" and "silent." Margaret W. Rossiter (1982), author of *Women Scientists in America: Struggles and Strategies to 1940*, for example, refers to the "invisibility" of early women scientists, invisible even to experienced historians of science, a result of a "camouflage intentionally placed over their presence" (p. xv). Mary Field Belenky and her colleagues wrote of the self-described powerless "silent women" in *Women's Ways of Knowing* (1986), and Paulo Freire spoke of Brazilian peasants as the "culture of silence" in *Pedagogy of the Oppressed* (1970). Like deaf men and women in history, these people without an affirmed political or intellectual voice have struggled for opportunity and dialogue. In *Silence of the Spheres* parallels can be seen between the struggles of deaf scientists and those of women and other marginalized groups. The various life-struggles of deaf men and women scientists are representative of the social and political issues faced by the "deaf population" as a whole; and, through the deaf scientists' own words, we learn much about the phenomenon of deafness as it relates to the pursuit of professional careers. Indeed, most of the life experiences of these men and women I examine in the physical sciences, medicine, engineering, and invention are generalizable to those of deaf social scientists, or deaf professionals in the arts.

In "The Gender Equation," Sue V. Rosser (1992) reviewed a number of writings on the underrepresentation of women in science. She wrote:

> The recognition that women may have a different perspective on science than men do opens the door for an understanding of other viewpoints that may come from members of different races, classes, cultural backgrounds and sexual orientation. Topics for research, theories and conclusions drawn from the data, and scientific practice may all suffer from biases other than sex. But again, such biases could be exposed and perhaps eliminated if there were scientists representing the diverse viewpoints. (p. 47)

Along the same vein, in his discussion of the benefits of participation of African-Americans in science, Gaston (1989) has written about how the choice of research problems in science has both cognitive and social bases. He examined the argument that it makes no difference who does science: "To advocate the notion that it does not matter who does science is to assume that sufficient numbers of other talented people will become scientists and therefore ensure its success" (p. 130).

Both in the physical sciences and the social sciences we have a rich variety of research which illustrates how the values and orientations of both deaf and hearing scientists come into play in their selection of topics and their methods of investigation. Ruth Benedict, one of the greatest anthropologists in history, never studied the cultural aspects of deafness, but as her friend Margaret Mead has explained, Benedict's focus in her anthropological fieldwork was indeed influenced by her deafness. In 1988, Carol Padden and Tom Humphries, both deaf, provided some excitingly unique insights in their book *Deaf in America: Voices from a Culture*. One of the most thorough and effective works written on the attitudes and prejudices faced by deaf people through history has been *When the Mind Hears: A History of the Deaf* by Harlan Lane, a hearing author. Similarly, when the physicist Robert H. Weitbrecht left the Stanford Research Institute to devote his time to the development and dissemination of the acoustic coupler, he was certainly influenced by his interest and need as a deaf man to communicate more effectively by telephone. Early experiments by deaf people such as Emerson Romero and continued persistence of members of the Deaf community also led to the research on the development of the television decoder technology. The issues of importance are underrepresentation and access to the same resources available to hearing scientists. Without doubt, the results of the scientific research conducted by deaf scientists in history have been of great benefit to humankind. Increasing the participation of deaf people in science will only expand these contributions.

Silence of the Spheres is a historical look at the interplay between science and deaf people. The book is not a history of the "Deaf community"; neither is it a history of science. Each of these areas has been studied extensively by others, and I have no intention of repeating such efforts. The convergence of these two, however, has never been examined. Of necessity, I repeat some facts that are common knowledge to enthusiasts in both domains, but I have also uncovered a plethora of stories and information not widely known. This historical review, however, focuses primarily on the contributions of deaf people to science, and of deaf scientists to society; it deals to a much lesser extent with the events precipitated by hearing people. To the latter belong such frequently reported historical events as Jerome Cardan's observation in the sixteenth century of a deaf man who could "hear" by reading and "speak" by writing, the establishment of the first schools for deaf children in Europe, and the infamous 1880 Edict of Milan in which hearing leaders at the International Congress on the Deaf paternalistically proclaimed the "superiority" of oralism over the use of sign language in the instruction of deaf children. More important to the theme of this book are the advances in science made by people who lived in silence while these events unfolded and the controversies raged on. I emphasize the continuously growing roles of deaf scientists in effecting positive change in the spheres of science and society.

All through history attitudes have taken many forms in presenting obstacles to men and women scientists who were deaf. The attitudes of some of the most influential thinkers, particularly Aristotle, delayed serious efforts to educate deaf children for centuries. Throughout this book, these barriers are interwoven. In Chapter 3, however, I focus exclusively on the confrontation of attitudinal barriers by deaf scientists in the latter decades of the nineteenth century and the early decades of the twentieth, without doubt one of the most disheartening periods in the history of deaf people in relation to the paternalism inherent in hearing society.

Amid the ongoing prejudice, discrimination, and patronization experienced by deaf scientists, certain movements and events in history have been factors which have provided these men and women with opportunity. In Chapter 1 I describe the rise of scientific societies during the Enlightenment and how this was a factor of importance to both congenitally and adventitiously deaf scientists. Chapter 2 briefly highlights the establishment in the nineteenth century of schools for deaf children in America, and the establishment of Gallaudet College, a second major historical thrust important to the theme of this book. In Chapter 4 the growth of industry and the onset of World War I are described as factors which opened new doors for deaf men and women in science, and in Chapter 5 the promise of higher education opportunities for deaf people interested in science and technology is presented, especially the National Technical Institute for the Deaf. Federal legislation and advances in technology which are rapidly improving the quality of life for deaf people are also described as important factors. In the Conclusion I discuss the need for greater roles and responsibilities to be taken by scientific societies in the modern era to effect positive change. I also describe some relevant issues which I believe, if attended to, would significantly increase the representation of deaf people (as well as persons with other disabilities) in the scientific community.

LEARNING FROM HISTORY

There is value in viewing history from different standpoints. In his 1988 essay in *Natural History*, the well-known Harvard University paleontologist and scientific writer Stephen Jay Gould stated that we can render different and interesting accounts from the standpoint of objects studied. We might have an engrossing account of astronomy, he wrote, if we could see the moon's point of view, or of genetics through the eyes of a fruit fly. The notion of examining scientific history from a nontraditional perspective is the underlying approach to this book. In *Silence of the Spheres* I describe the standpoints of human beings who for too long have been treated as "objects" or "defective" human beings. It is a glimpse into the history of science—based on the perspectives and life experiences of deaf persons. Through this effort I hope to not only reveal the accomplishments of some deaf men and women in science, but also to help break down the attitudinal barriers which they continue to face.

To a deaf scientist like myself, and to many readers who enjoy looking at history in different ways, it becomes at once curious and romantic to reflect on how deaf men and women have contributed in the scientific communities of their times. Deafness often consumes those who experience it and metamorphoses their lives, yet this natural human condition has not previously been used as a window through which one views scientific history. All of us, hearing or deaf, have been moved to wonder and imagine as we read the stories about Isaac Newton and the falling apple, or Galileo and the Inquisition. The events leading up to the discovery of penicillin and the invention of the telephone have often been told, each time the writers calling to our minds images of perseverance, genius, and other qualities which inspire us toward our own goals, however lofty or modest they may be.

Educators today tell us that deaf children have trouble reading. Texts describe deaf children as shy; others assign deaf students behavior problems. One is inclined to believe that deaf children are poor students in general. But, then, history has shown us that some of the greatest accomplishments in science have been attributed to men and women who, in their childhood, exhibited similar characteristics, as perceived by their teachers and parents. Albert Einstein and Wernher von Braun had great difficulty with mathematics in the early grades. Henry Ford had trouble reading and Isaac Newton was a poor student in general. Henry Cavendish, excessively shy even in adulthood, almost never spoke, and communicated with his servant through notes. Coincidentally, the man who brought Cavendish's work in electricity to publication one hundred years later, James Clerk Maxwell, had a solitary childhood in the country which made him extraordinarily reticent as well. Orville Wright was expelled from school for behavior problems. Thomas Alva Edison, before he lost his hearing, was considered "addled" by his teacher, and his parents took him out of school to educate him at home.[2] These are but a few of many such stories which lead us to realize that there should be rich returns in understanding better the role attitudes play. The barrier created by attitudes about deafness and deaf people, in particular, has revealed itself in many forms through history, often out of ignorance. Fundamental to this ignorance is a general lack of awareness about deaf people who have been successful in history. Such a need for more information about scientists with various disabilities is being increasingly recognized on the national level. Textbook publishers such as Addison-Wesley Publishing Company, for example, are seeking such information to make science more accessible to all students, and the National Science Foundation Task Force on Persons With Disabilities has described as a major issue the "virtual invisibility of role models in science, science education, and engineering for children with disabilties" (National Science Foundation, 1990, p. iv).

My goal will be met if this book can take the loss of a physical sense which may be, for some readers, perceived as a pathological condition, and transform that perception, through the stories related, into an understanding of a complex human experience worthy of personal reflection and scholarly research. Indeed, the information I have collected for *Silence of the Spheres* only reinforces what I have learned from my conversation with Stephen Hawking in Boston: Human perspectives about a disability are in the eye of the beholder. And these perspectives should certainly be a part of the history of science.

THE DEAF SCIENTIST: DEFINING THE REALITIES

"The deaf" are described in historical and contemporary literature in much the same way as Lord Florey once explained "scientists" in a lecture in Melbourne in 1963: "enumerated, divided into categories, constructed into tables, illustrated by graphs, and pronounced upon in bulk. But it is sometimes forgotten that they are human beings." In searching for an appropriate way to identify scientists who were, or are, "deaf," I have chosen Jerome D. Schein's functional definition of "deaf people" as those who generally "cannot carry on a conversation with their eyes closed" (Schein, 1975, p. 18). This definition includes a strong reference to communication, one of the most important processes of science, and would therefore include men and women with hearing losses presently estimated at 70 decibels or more.

But there are multiple realities to consider in defining the "deaf scientist" that are not immediately apparent on a first reading of Schein's functional definition. Two such realities are the age of onset and the amount of hearing loss. The very first deaf American scientist to earn a Ph.D. from a university met with this categorical fate, one that would be repeated again and again with other scientists through the years. After graduating from Yale in 1861, Gideon E. Moore entered the University of Heidelberg in Germany for further study and earned a Ph.D. in 1869. Following this there appeared editorial comments and arguments about Moore's deafness in the major journals of the American Deaf community. On December 24, 1870, a *New York Tribune* reporter described Moore as entirely deaf, thus precipitating several remarks by editors in the Deaf community, including Edward Allen Fay of the *American Annals of the Deaf and Dumb*, Edwin Hodgson of *The Deaf Mutes' Journal*, and Weston Jenkins of *The Silent Worker*. Jenkins argued that Moore could hardly be considered a "deaf-mute" since he was fourteen years old when he lost his hearing and "at that age he was already a promising student in literary and scientific branches" (*SW*, *3*[23], 1890, p. 3). When others claimed that Moore had a progressive loss of hearing and was not entirely deaf, Jenkins confirmed that Moore had "lost his hearing in childhood, not suddenly but by degrees," and Hodgson, a close friend of the chemist, stressed that Moore had reached total deafness in adolescence (Braddock, 1975, p. 35). The age of onset of deafness and the degree of hearing loss, then, were considered important factors, not only in terms of advantages in speech and spoken language development, but also in relation to how a scientist may be perceived as belonging to the Deaf community. By and large, there is one lesson we learn—a scientist can be "functionally deaf" by having a hearing loss, but may not necessarily be "culturally deaf" (that is, having an identity in the society of deaf people). Such a discussion of sociology is, of course, beyond the scope of this present work and I refer the reader to an excellent book entitled *Identity Crisis in Deafness* (1979) by Benjamin M. Schowe.

It is interesting to examine how the age of onset of deafness has been perceived through history. At least up to the early twentieth century, it was the perspective of many people that a congenitally deaf person is much less likely to attain a quality formal education as compared to one who became deaf adventitiously. While adventitiously deaf people have the benefit of some experience with hearing speech and spoken language, mistaken generalizations are often made about the ability of children deafened before the age of five or six to learn "language" effectively. First, as this book will reveal, congenital and early-onset deafness does not inevitably inhibit mastery of English language and subsequent success in science, particularly when an adequate education with minimal attitude barriers is provided. Second, it is unfortunate that the mastery of American Sign Language as a *visual* language is often not recognized, and not capitalized on (1) to further develop English as a second language, and (2) to communicate science and other subjects effectively through the visual modality.

In the anecdotal reports found in the literature of the Deaf community during the nineteenth century, especially prior to the establishment of the National Deaf-Mute College (now called Gallaudet University), editors brought attention to the achievement of any deaf person in science or mathematics, regardless of age of onset, because such attainment was considered unusual. Harvey Peet, for example, wrote in his "Report on the Education of the Deaf and Dumb in the Higher Branches of Learning" that "there has been an instance in France in which a deaf man (Paul de Vigan) went through a course of the Physical Sciences with distinction" (*AAD*, *5*[1], 1852, p. 57). Peet also mentioned in this report that there were "semi-mutes [adventitiously deafened] both in Europe and in America (Dr. Kitto of London, and James Nack of New York) who have attained to eminence as scholars, and have been successful in authorship." Six years later, a story appeared in the *Philadelphia Saturday Evening Post* about Johannes Michel Moser, a "prodigy in arithmetic," and a native of the old town of Ratisbon, in Bavaria, who had moved to the city of Brest in France. Moser, born deaf, was described in the *Post* as "an arithmetician of the first order, executing the most complex and difficult numerical calculations with the rapidity and precision of a calculating machine" (*AAD*, *10*[3], 1858, p. 189). In 1910, a civil engineering student, Herr Poulsen, was held up proudly as the "first born-deaf Dane successfully to be prepared for tertiary education," passing an examination which gave him admission to a university at Roeskilde, Denmark (*VR*, *12*[9], 1910, p. 582). And in 1917, another deaf man, William Naglo, was highlighted in a popular book as having "distinguished [himself] as a master of electrical science" (Roe, 1917, p. 135). After studying in England, Naglo established a business in Berlin and assisted in laying one of the transatlantic cables.

These success stories were so few in number during the last half of the nineteenth century and the first decades of the twentieth century that they were considered newsworthy enough to be reprinted in local periodicals around the country. The importance of such recognition in the literature of the Deaf communities is best summarized by a reporter in the *British Deaf Times* in 1910 who, in describing the accomplishments of a Mr. Maddison attending the Royal College of Science in South Kensington, England, wrote that "too much publicity can not [sic] be given to these instances of successful deaf, in order to counteract the absurd prejudice against them as a class" (*SW*, *22*[9], 1910, p. 177).

Data on the age of onset of deafness among scientists are not easy to find. Contemporary references which include ages of onset of deaf scientists include the American Association for the Advancement of Science *Resource Directory of Scientists and Engineers with Disabilities* (Stern, Lifton, & Malcom, 1987) and the *Gallaudet University Career Information Registry of Hearing Impaired Persons in Professional, Technical, and Managerial Occupations* (National Information Center on Deafness, 1987). Neither claims to provide data representative of deaf persons in the United States who have held professional positions in the sciences. These references, however, do illustrate that deaf individuals of any age of onset can succeed in scientific careers. In addition, for the 246 deaf persons with graduate degrees in science for whom I have data on age of onset of deafness, 62 percent were either born deaf or lost their hearing in their first five years (59 percent of 205 men and 78 percent of 41 women).

A third reality in defining the "deaf scientist" is involvement in the Deaf community. Through an analysis of the literature, we learn about how some successful deaf scientists have made deliberate efforts to develop a sense of belonging to the Deaf community, while others have not. For a deaf person to be accepted into the Deaf community, involvement with other deaf people is essential. So was and is learning to communicate in sign language. Thirty years after Gideon E. Moore left Yale, the editor Weston Jenkins commented that "from his long and intimate association, both at home and at school, with his deaf-mute [sic] brother . . . [Gideon E. Moore] has become an adept [sic] in the sign language and he has a great deal of interest in and sympathy with deaf-mutes [sic] as a class" (*SW*, *3*[23], 1890, p. 3). In this note, made on the occasion of Moore's visit to a school for deaf children to relate stories about his European travels, mention of his deaf brother, H. Humphrey Moore, an "eminent artist," is important. In referring to another deaf member of the Moore family, especially one who had established himself in the Deaf community, Jenkins likely increased the chances for Gideon to be accepted as well. Other such reports about deaf scientists underscore this notion of acceptance into the Deaf community. A prominent nineteenth-century deaf dentist, George K. Andree, for example, was appointed to serve on the Board of Dental Examiners of Oklahoma and was elected president of the Oklahoma Dental Association. Even though he graduated from Gallaudet College in 1902, his apparent lack of involvement with other deaf people may have prompted the following comment to be published by a member of the Deaf community: "Mr. Andree does not mingle with the deaf as far as we know of, so personal congratulations are out of the question; but if he should chance to read these few lines, he will know we are glad of the honor that has come to him."[3] Contrast this with the report about the "almost totally deaf" physician Horatio Brewer, a veteran of the Civil War who, although late-deafened, was welcomed publicly to the Pas-a-Pas Club. Commending Brewer for his determination to join the Deaf community, the editor wrote that it "is a novelty to see a man of Dr. Brewer's profession a member of such an organization. He is making good progress at mastering our signs and as he is quite a lecturer and public speaker the club's 'hit' will have an added attraction 'ere long" (*SW*, *23*[2], 1910, p. 29).

Historically, the diversity of attitudes of deaf scientists about their participation in society is no different than that of scientists with normal hearing. As with hearing scientists, there have been deaf men and women whose careers were fabrics woven of contributions to both science and society. There also have been those whose scientific interests consumed their lives. The men and women who were advocates for the rights of deaf people were much easier to find in the literature of the Deaf community. The nineteenth-century horticulturalist George W. Veditz established a floral business in Colorado. He was the editor of several journals related to plants and flowers and set up the first Peony and Iris Show in Colorado. Veditz also held the office of president of the National Association of the Deaf and was a brilliant advocate for deaf citizens in the United States. Gerald M. McCarthy, the state botanist and entomologist of North Carolina in the 1890s, was a graduate of Gallaudet College who took time aside from his scientific studies to challenge Gallaudet administrators on the issue of exclusion of deaf persons from the normal school for training teachers.

This discussion may seem strange to those unfamiliar with the subject. There is implied an actual fragmentation of the Deaf community by age of onset, amount of hearing loss, and sociological factors. The tendency to classify people with hearing losses into discrete categories has resulted in a measure of cultural inclusion or exclusion. But as a "functionally deaf" person, every scientist in this book has shared many of the same experiences of full-fledged members of the Deaf community, especially the constant challenge of gaining access to information.

Some readers may wish to know what is meant by "Deaf culture." To define this term, I have chosen the words of a distinguished anthropologist who was herself confronted with the communication and attitude barriers associated with deafness throughout her life. "A culture, like an individual," wrote Ruth Benedict (1934), "is a more or less consistent pattern of thought and action. Within each culture there come into being characteristic purposes not necessarily shared by other types of society" (p. 46). Although it was not Benedict's intention, this definition, in my opinion, is a good one for describing "Deaf culture." The "consistent pattern of thought and action" is seen in the contributions of deaf people, their languages of signs, their art and poetry, and other aspects of their heritage. The culture of deaf persons in America, in particular, has been the focus of hundreds of articles and books, and these various analyses and compilations of facts and anecdotes reveal the heritage and values of people who have come together as a community primarily because of the shared experience of deafness. Jack R. Gannon's *Deaf Heritage* (1981) is an extensive collection of facts and anecdotes about deaf people in American history which portrays this pattern in a most revealing way. The three-volume *Gallaudet Encyclopedia of Deaf People and Deafness* (1987), edited by John V. Van Cleve, includes summaries of various aspects of the Deaf community by many deaf and hearing professionals. These three volumes provide evidence that deafness is a cultural rather than pathological experience for many deaf persons. Other works such as *Deaf in America: Voices from a Culture* (Padden & Humphries, 1988), *Great Deaf Americans* (Panara & Panara, 1983), and *Notable Deaf Persons* (Braddock, 1975) are rich sources of information. Videotapes such as Loy Golladay's "Off-Hand Tales" have helped to preserve the visual beauty of American Sign Language (ASL). Various anthologies have brought recognition to deaf writers of the past. Robert F. Panara's (1954, 1970, 1974) analyses of deaf writers and poets, for example, demonstrated the cultural aspects of deafness and the interplay with the society of hearing people. The deaf anthropologist Simon Carmel has studied the recollections of deaf survivors of the Holocaust and the insights and resources of the Jewish deaf population (Levine, 1991). He has also published work on "deaf folklore" (Carmel, 1987). In this variety of ethnographic work, however, is missing a substantive analysis of the contributions of deaf scientists, as well as the impact of science on the lives of deaf individuals and communities.

A fourth reality in defining the "deaf scientist" is that of self-identity. While many deaf people have elected to view their world of silence as a cultural phenomenon, others have not. In part, a disregard for the cultural aspects of deafness may be due to the presence of certain elements which Deaf culture does not share with many other cultures. One aspect, in particular, is the lack of family identity for most deaf people. While parents may pass down the heritage of African-Americans to their children, for example, most deaf children have hearing parents, and many never meet a deaf adult through their formative years. In addition, much of the educational literature on deafness has focused on a pathological perspective of sensory loss. This view has historically stigmatized deaf people. Some scientists have felt the stigma associated with hearing loss to an extent that they do not identify themselves as "deaf" in *any* sense of the term.

With these realities in mind, the reader will understand that it is not a simple matter, as it is with gender or race, to determine who constitutes the historical and rather small population we may refer to as "deaf scientists." But, importantly, it was not my intention to limit my search to those men and women who are culturally deaf. This would do injustice to many accomplished scientists, and present an inaccurate portrayal of the contributions and successes in history and in contemporary society. A few of the scientists in this book, it may be argued, were "hard-of-hearing" rather than "deaf," but self-reports, and the descriptions of their associates, indicate that they may well have been "functionally deaf" in their times, especially before effective amplification became available. *Functional* deafness, then, has been the primary criterion for inclusion in this book, and this permits me to examine the lives and work of men and women with a wide range of interests and needs.

DEAFNESS: A DIVERSITY OF EXPERIENCES

Although technology was insufficient in earlier times to accurately measure the amounts of hearing loss experienced by many scientists and though the word "deaf" may have been used more loosely, the anecdotes and personal writings of the scientists included in this book have been helpful in determining the extent of their deafness. More important, it becomes evident that deafness has not stopped them from contributing to nearly every sphere of scientific endeavor. There are some common threads which can be found in the stories of their lives and work. For many deaf scientists it has been a case of living in two worlds, sometimes a feeling of not belonging to either, and the diversity of experiences these people have had presents a moving panorama of the struggles for opportunity and access.

First, the biographies, diaries, personal journals, and other reports of deaf scientists in history emphasize the importance of parental support during childhood. Witt and Ogden (1981) wrote that parents can exercise a powerful influence on the motivation of a deaf child:

> To a very great degree, young children derive their sense of their personal value from their parents. In families where the parents are accepting, nurturing, openly loving, and encouraging, chances are good that children will develop sound self-esteem. But children raised by cold, negative, and overly critical parents are in danger of growing into adults whose estimation of their own personal value is very low. For the child with

> physical disabilities, the risk to self-esteem is compounded. Working against the development of a strong sense of personal worth are three complex and powerful forces: the child's own sense of being 'different'; the parent's difficulties in accepting and adjusting to the disability; and the negative reactions handicapped [sic] people meet with in the world at large. (p. 5)

These authors explain how deaf people have been viewed "with a mixture of fear, scorn, distaste, misunderstanding, and pity" throughout history (p. 6). Parents must adjust to the shock of having a child who is deaf, and to the separation between them which may occur because of the deafness. In addition, there are the responsibilities associated with raising a child with special needs, as well as feelings of guilt which some parents experience. As Witt and Ogden summarized, responses of parents range from loving support and healthy relationships to hypercriticism, or from overprotectionism to abandonment and a sense of defeat.

When schools for deaf students began to open in the United States in the early nineteenth century, parents responded by placing their children in these programs. Through the years, parents also helped to found schools, financially supported educators in establishing them, and provided assistance to scientists and inventors who were interested in advancing the quality of life for deaf people. The story of Charles Shirreff, whose father's concern led to the establishment of the Braidwood Academy in Great Britain, was perhaps an impetus for Francis Green, whose deaf son met with success at the same academy, to begin a similar school in the United States. Similarly, the physician Mason Fitch Cogswell, whose daughter Alice was deaf, encouraged Thomas Hopkins Gallaudet to travel to Europe to learn more about the successful methods of teaching deaf children, and Gardiner Greene Hubbard, also the father of a young deaf woman, provided financial backing for Alexander Graham Bell's teaching and scientific studies in Boston.

By and large, the more successful deaf scientists have case histories in which parental support and a consequent higher self-esteem have been characteristic. "I owe to my mother," the nineteenth-century fossil hunter Leo Lesquereux wrote in gratitude, "all the influences which were developed later in my youth. Until I left home she directed my studies at school, and after I left it I never ceased to correspond with her until her death" (Lesley, 1890, p. 190). The stories of Lesquereux's forays into the woods, and his mishaps, suggest a mother remarkably free from overprotectiveness. Similarly, the deaf Russian rocket pioneer Konstantin Tsiolkovsky's mother taught him to read and write. Although she died when he was thirteen years old, she succeeded in inspiring in him a love for life. The eighteenth-century astronomer John Goodricke's parents sought out the Braidwood Academy at a time when it was the only school for deaf pupils in Great Britain. In this same vein, the twentieth-century deaf bacteriologist Harold J. Conn, in a tribute to his father, Herbert William Conn, also a bacteriologist, provided a detailed retrospective analysis of their relationship:

> Meanwhile [my father] was not neglecting [my] education, either, although his ways of directing it were so unobtrusive I did not recognize them as such

> at the time. If I were writing a story of my own life instead of his, I would have much to say on the subject; but in one particular his method of molding my life had such a bearing on his own personality that it must be included here. I know it was a great blow to both my parents to realize—as was only too evident by this time—that my hearing was failing and that doctors were helpless to save it. But never a word did I get from them to indicate how they felt; and I know their reticence was due to Father's belief that nothing must reach me that might destroy my confidence in my ability to succeed in spite of such a difficulty. So successful was he in this that not until life had given me a few hard knocks did I realize that it might be harder for a handicapped person to succeed than for one with normal faculties. Perhaps such knocks were harder than they would have been if I had been taught to expect them; but I have never regretted that I was brought up in such a way as to feel that my deafness was never an excuse for failure. Temporarily I might lose self-confidence after some unpleasant experience, but the tendency to be self-reliant, in spite of my difficulty, was so ingrained that sooner or later I always recovered my confidence. I am sure that whatever success I have achieved, even though handicapped in some respects, is due in large part to his attitude toward me while I was a boy.[4]

A second common characteristic that emerges in the review of the lives and work of these deaf men and women is the struggle for access to information and dialogue. Deafness is sometimes referred to as a "communication handicap" and in educational and occupational environments it often remains one. In one study, Gallaudet University's National Information Center on Deafness (1987) surveyed deaf professionals to identify problems they encountered in furthering their education or career. Communication was mentioned in 70 percent of the 366 problems described by the 216 respondents. The next frequently cited problem was employer discrimination (19 percent of the respondents), followed by knowledge or educational deficiency (6 percent).

The stigma associated with the lack of voice quality in a world in which spoken communication is dominant is a greater handicap for some than the loss of hearing. People unable to hear their own voices have difficulty monitoring volume and articulation. Even in the case of adventitious deafness, it is common for one's speech to lose clarity. For some scientists, a feeling of marginalization associated with less-than-normal speech and hearing may lead to an avoidance of public lectures and teaching. Most of the deaf scientists who have lectured in professional societies and at universities became deaf adventitiously. On the other hand, even with normal speech, the twentieth-century American zoologist Hubert Lyman Clark, who lost much of his hearing while a young adult, found the communicative demands inherent in lectures to be too much a burden. But, in contrast, and despite the challenges in communicating with hearing persons, many individuals born deaf or with early-onset deafness have accepted a variety of teaching assignments in university settings. Regardless of the means these deaf scientists have employed to communicate (speaking, using voice interpreters, writing, etc.), persistence has been the key to success.

A wide range of strategies for overcoming the communication barrier imposed by a hearing loss within the context of dominantly spoken communication environments is also evident in the biographical materials.

Some strategies were clever, others humorous, but all were fairly effective. The common pad and pencil is mentioned repeatedly, and one cannot underestimate the power of these tools. As a member of the German parliament in the 1870s, the profoundly deaf social scientist Heinrich von Treitschke sat next to the shorthand writers and once he had learned the abbreviations he was able to participate fully in the debates. Thomas Alva Edison fooled every member of the naval consulting board with his ability to communicate, including President Woodrow Wilson and Secretary of the Navy Josephus Daniels. As one reporter described the incident, Edison "heard little that was said, but he presided successfully because his assistant, Mr. Miller R. Hutchinson, kept him informed of everything by means of a telegraphing finger tip that touched Mr. Edison's knee under the table" (*SW*, *28*[10], 1916, p. 199). And such ingenuity finds new forms through time. Capitalizing on technology to bridge the communication barrier, J. Tilak Ratnanather, a British mathematician who earned his Ph.D. at the University of Oxford, participates in the research community on an equal basis with his hearing peers through the use of electronic mail, and facsimile and interactive video conversations, and he has presented papers on this method of communication to deaf and hearing scientists.

Adaptability, perseverance, a sense of humor, and an ability to communicate in whatever way is successful have been widespread strategies which proved especially valuable during the years these deaf men and women scientists spent in university settings. Michael Weiner, born deaf from maternal rubella, began his education at the Pennsylvania School for the Deaf and later graduated from Temple University School of Dentistry. Weiner sat in the front of the room and took his own notes in many of his courses (*Philadelphia Evening Bulletin*, September 26, 1975). Steven Rattner did not benefit much from the slide presentations in the darkened rooms, but after the lectures were over, he found the slides in the library along with written explanations. Born deaf, Rattner learned both sign language and speechreading, and when communication with a patient was difficult during his dental work, he read the lips of his assistant standing behind the patient (Moore, 1987).

Long before hearing aids became available, deaf scientists found means for using whatever residual hearing they may have possessed—through speaking trumpets, mechanical contraptions, "flexible conversation tubes," and other inventions of their own. Nowadays, in addition to electronic mail, deaf scientists have access to captioning decoders, improved hearing aids, visual signaling devices, telephone relay services, and other services which facilitate communication between deaf and hearing callers by translating back and forth signed, spoken, and typed messages.

Another effect of hearing loss common to most of the deaf scientists discussed in this book is social isolation.[5] Stanley E. Willis (1959), an engineer and technical editor who was deafened while in England's Royal Air Force during World War II, summarized his experience poignantly as follows:

> I know, only too well, that single and most damaging effect of deafness, the feeling of loneliness it creates. My social contacts were few and were neither rich nor satisfying. Because I could not hear well, I felt cut off and lived in a world of silence—a world shrouded with doubts, fears, and

> suspicions. Further and further I seemed to crawl into myself, and the vicious cycle continued. (p. 4)

While investigating materials in the libraries of Europe and the United States, one question lingered in my mind: Was deafness a hindrance or a help to the scientists who reached prominence? Their autobiographical comments lead me to believe both interpretations. Some deaf men and women have attributed their commanding positions in the world of science to the increased powers of concentration afforded them by virtue of their loss of hearing. Those who fully appreciated the advantages of silence were quick to credit their deafness for the ingenuity and industry with which they approached their work. The seventeenth-century physicist Guillaume Amontons, as well as the inventor Thomas Edison, actually refused medical assistance because they had learned to appreciate the isolation and solitude afforded them as a result of deafness. The twentieth-century paleontologist Tilly Edinger often switched off her hearing aid when she wished to concentrate on her work. But these were by no means common feelings. While being physically or sensorially different is often stigmatic, deaf men and women have clearly achieved preeminence in science—whether they were or were not bothered by their deafness. In *Silence of the Spheres*, I therefore describe how deaf scientists have reacted similarly and differently to a loss of hearing, neither to praise those who have assimilated into the societies of deaf and hearing people, nor to praise those who have not, but to illustrate the range of influences and options that deafness has presented them.

Without doubt, the technological and human services now available, as well as the continued application of the inventive talents of deaf people, will help to some extent in providing improved access to communication and in reducing the isolating effects of hearing loss. Yet, there are no better solutions, as these men and women will attest, than awareness and sensitivity among hearing educators, employers, and others with whom they work. Thomas H. Huxley once said, science is nothing but "trained and organized common sense." The same may be said for the accommodation of people who are deaf.

1

The Enlightenment and the Rise of Scientific Societies

If "history is the essence of innumerable biographies" as Thomas Carlyle once wrote, then there is no history of deaf people during antiquity and the Dark Ages. Those in search of information about deaf people living in the period up to the Renaissance are left with a meager collection. Everywhere in pre-Renaissance history we find mythical and magical references to deafness and the human destiny of deaf people denied them in their struggle against oppression and persecution. Shunned as unfit to learn by such influential thinkers as the philosopher Aristotle and the poet Lucretius, rarely are deaf people mentioned by name.[1] Here and there one may read an occasional story about deaf children of wealthy Roman consuls who had taken up a trade, usually in art, but not a single report about a deaf student of natural philosophy or science. Then came the tenth-century physician of Chartres "John the Deaf," teacher of Droco in the period known as the Carolingian Renaissance during the "Age of Alcuin." Next came the thirteenth-century Dominican monk "John the Deaf," who was interested in optics, summarized Aristotelian physics, and published on meteorology and the rainbow. But absolutely nothing can be found about their deafness in the extant literature, and it may be possible that "Le Surdus" (i.e., "The Deaf") was a family name. The obscurity surrounding the lives of these two men casts doubt on the distinction they might hold as the first deaf scholars in world history.[2]

Not until the mid-sixteenth century do we find the first mention of the ability of deaf persons to read and write, this provided by Rudolph Agricola, a Dutch humanist. Jerome Cardan, a mathematician and physician from the University of Padua in Italy, himself with a son deaf in one ear, gave further visibility to Agricola's observation in his own book titled *Paralipomenon*, published posthumously in 1663. Nearly every history of the education of deaf people has recognized this brief observation as an important milestone. Davis and Silverman (1978), for example, have referred to Cardan's writing as the "Magna Carta for the Deaf" (p. 423), and Bender (1981) called it a "revolutionary declaration," for "it broke down the long-established belief that the hearing of words was necessary for the understanding of ideas. It recognized the ability of the deaf to use reason" (p. 30).[3] However, great thinkers as early as Socrates (*Cratylus* in Plato) four hundred years before the time of Christ and as recent as Leonardo da Vinci in 1499 (*Codice Atlantico*) had observed such ability to reason among deaf people through gestures and signs. But it was not until the sixteenth century that the stigma associated with deafness began to lift, albeit slightly, with the religious rejection of deaf people challenged by the writings of the clergy, the administering of communion to them, and the first efforts to teach them. And, importantly, the appearance of respected deaf artists and writers in this period brought further recognition that deaf people were capable of reasoning, of learning language through the sense of sight, and of contributing to society.[4]

At about this time, occasional reports of congenitally and adventitiously deaf people learning science also began to appear in the literature. Early records indicated that some congenitally deaf persons had learned mathematics and science from the world's first teacher of deaf children, Pedro Ponce de León, in the late 1500s. The first deaf teacher of France for whom there exist any records was himself a prodigy in science and mathematics. He was Étienne de Fay, born in 1670, one of the earliest congenitally deaf pupils to be recognized for an education in science which he applied in his subsequent work as an instructor. Father Cazeaux, prior of the Abbey of Notre Dame de Beaumont, speaking to the Royal Academy of Belles-Lettres of Caen, explained how he had sent the orphan to a school for deaf children around Amiens. Little is known about how the "White Friars" were so successful with de Fay, who went on to become a teacher at the abbey himself. Later, Father André visited de Fay's school and referred to the deaf teacher as a "mathematician, a man of learning, an architect, a sculptur; in short, a man of universal knowledge" (Braddock, 1975, p. 40).

When deafness occurs later, after scientific interests have been established, changes in the patterns of scientific study may be as necessary as adjustments in one's social life. One may at first be inclined to dismiss late-onset deafness with the attitude that it is a common phenomenon. I would argue with this. In *The Early Naturalists: Their Lives and Work (1530–1789)*, for example, L. C. Miall (1912) briefly mentioned Charles Butler, a British naturalist who lived in the last decades of the 1500s and the early decades of the 1600s. Butler was forced to discontinue his work on the sounds made by bees because of his deafness. There is no explanation of the extent of Butler's hearing loss, or at what age the hearing loss began, but in the discussion of his work it becomes obvious that he was deafened after he had begun his scientific studies. Butler went on to publish *The Feminine Monarchie; Or a Treatise Concerning Bees and the Due Ordering of Them* in 1609. He was a parson of Laurence Wotton, near Basingstoke, and also wrote on such subjects as rhetoric, consanguinity in marriage, and English grammar, as well as a book on the principles of spelling. The detail in which Butler pursued his earlier work with the sounds of bees leaves us to wonder whether his experience with deafness was as fascinating, and disturbing, to him as Beethoven's was in the history of music.[5]

More than any other factor, the rise of scientific societies marked the emergence of a first generation of deaf scientists. It is common to trace the lineage of scientific societies back to Plato, and even in the literary records associated with Plato's famous Academy we find a Socratic discussion of the ability of deaf persons to communicate with gestures.[6] As scientific societies spread through Europe, particularly in Naples, Rome, Leipzig, and Florence in the sixteenth and seventeenth centuries, they became centers where experimentation was nurtured, including investigations of deafness. In the early reports of these academies, we find studies on the anatomy of the ear and the use of tubes and trumpets for improving hearing. Then in 1662 came the London Royal Society and in 1666 the Royal Academy of Sciences in France, both providing constant and invaluable contributions to the study of deafness and serving as rich sources of information about the few deaf scientists who lived in the seventeenth and eighteenth centuries. The publications of these English and French academies provide us with a unique opportunity to examine how members who were deaf themselves, as well as deaf persons who were not members, were able to contribute to the advancement of science during the Enlightenment.

FRANCE AND THE ROYAL ACADEMY OF SCIENCES

Readers interested in the "deaf experience" in the history of science will find the seventeenth and eighteenth centuries a fascinating period in which eminent men of science were involved in examining the efficacy of instruction of deaf pupils. Among them were Royal Academy members such as the philosopher Jean-Jacques Rousseau, instigator of the French Revolution; his compatriot Denis Diderot; and the naturalist Georges Louis Leclerc, comte de Buffon, keeper of the Jardin du Roi and author of the forty-four volume *Natural History*. For some of these great thinkers, personal experiences with deafness may have helped to shape the perspectives evident in their scientific and philosophical writings (Rousseau and Charles de la Condamine, for example).[7] For others, interest was piqued by rather mundane observations of deafness, such as M. Felibien's description of how able diviners had sought to understand "a singular event, perhaps never before heard of, that hath just happened at Chartres. A young man between 23 and 24 years old, a tradesman's son, deaf and dumb [sic] from birth, began all of a sudden to speak, to the great amazement of the whole town." Medical demonstrations associated with deafness also can be found in the early Academy records. One can imagine these distinguished men of science at the demonstration in 1718 of Father Sébastien Truchet, a Carmelite monk and member of the Academy, who presented an acoustic eardrum made of ox gut stretched over a hoop and explained how two such drums could be fastened to the ears of a deaf man and their apertures turned toward the mouth of the speaker.

Deaf Members of the Academy

While deafness was a subject of interest, much less visible in the early history of the Academy were the deaf scientists who silently made their contributions. Such was the case of Guillaume Amontons (1663–1705), the first deaf physicist in history, who followed the impetus of Galileo and other

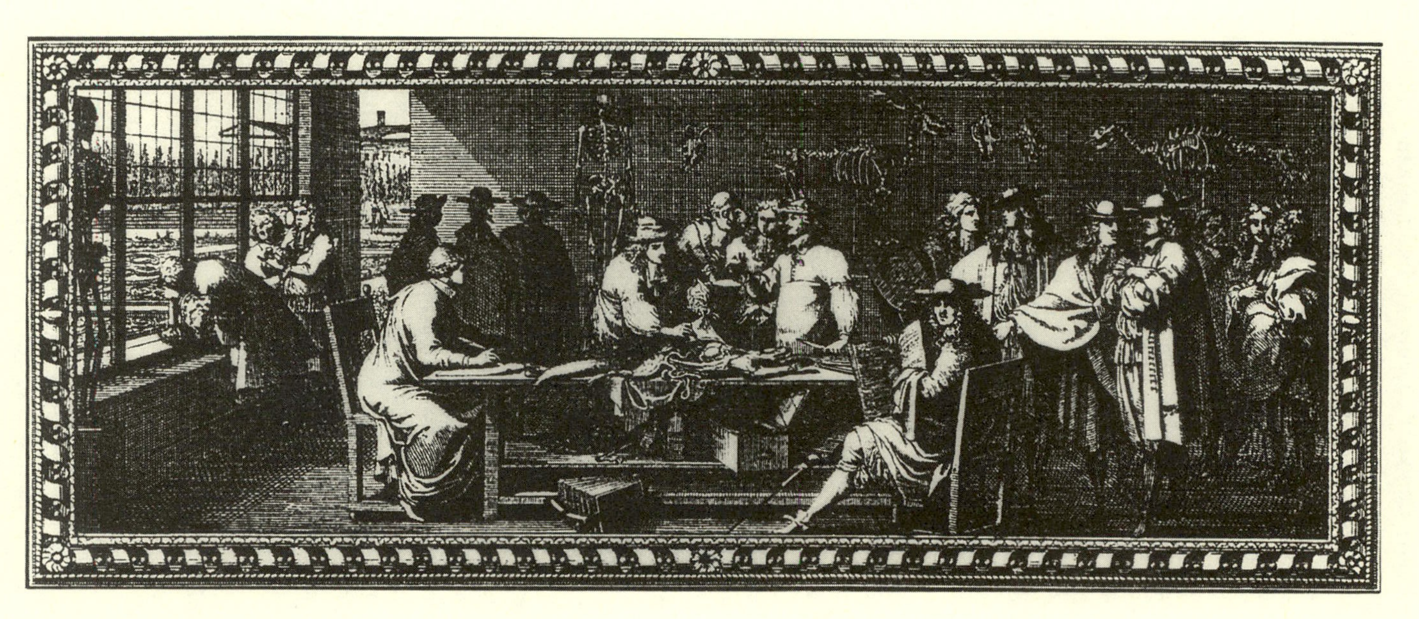

The French Academy of Sciences in the Eighteenth Century

scientists in constructing various instruments to study heat and temperature. Amontons is first mentioned in the Academy records in 1684, when he demonstrated a new hygrometer. As one of the most active of "les Académiciens," he was able to concentrate all the more because of his profound deafness. In Amontons' obituary in the *Memoirs* of that year, Bernard le Bovier de Fontenelle, a French science writer and the secretary of the Academy, described the onset of the physicist's deafness: "He was still in third grade when he became seriously deaf as a result of an illness, which sequestered him almost entirely from interactions, at least useful interactions, with others. Being all by himself, and given to thoughts which came from the depths of nature, he began to dream about machines" (p. 150). Amontons did not view his deafness as an affliction, refusing to be treated by physicians. As historian Florian Cajori (1962) wrote, Amontons' loss of hearing permitted him to pursue his scientific endeavors with "less molestation from the outer world" (p. 113).

Amontons was one of the first physicists to describe the phenomenon of a liquid cooling when it is evaporating, a principle which was to have significant applications. In a demonstration to the Academy, his hygrometer was used to illustrate how changes in humidity produced contraction and expansion of a small sphere of wood or leather. These contractions, in turn, caused the rise and fall of a liquid in a tube. Thus, he helped to lay the foundation for subsequent study of the atmosphere. His hygrometer was later perfected by DeLuc. Amontons' work went far beyond meteorology, however. He stimulated Fahrenheit to study the construction of thermometers and, as Cajori (1962) has written, "It is an interesting fact that Amontons's researches amount to an experimental proof of the law of gases now named after [Jacques] Charles and [Joseph Louis] Gay-Lussac, and that he first arrived at the notion of absolute temperature" (p. 114).

Amontons also devised a system for long-distance communication through optical telegraphy that he demonstrated to the royal family. He published numerous reports in the Academy *Memoirs* on thermic motors, the establishment of heat in a prism, and a confirmation of Leonardo da Vinci's estimate of the force needed to turn a crank. Death cut short this remarkable scientist's work at the age of forty-two.

Another member of the Royal Academy of Sciences at the time of Amontons was Joseph Sauveur (1653–1716) whose accomplishments in the field of acoustics are extraordinary considering his impaired auditory perception abilities. As V. V. Raman (1973) has written, "One of the ironies of human history is that little Joseph, who was to contribute much to the study of the science of sound, was born with defective hearing" (p. 161). W. S. Pratt, in *The History of Music* (1935, p. 325), also wrote of Sauveur's deafness: "[Sauveur's] work is astonishing, since he was a deaf-mute [sic] from infancy, hearing never but a little and acquiring but a partial use of his voice. . . . From about 1696, with the aid of musical assistants, Sauveur became absorbed in acoustics, joined the Académie, and published in its Transactions several epoch-making studies (1700–13)."

Sauveur's work in acoustics was preceded by others during the early years of the Royal Academy. The mathematician Marin Mersenne, for example, examined auditory perceptions, exchanging views with Jean Baptiste van Helmont and René Descartes, and undertook a thorough study of light and sound, vision and hearing, and the languages of man and animals. This interest in acoustics also led Mersenne to examine issues in the education of people who were deaf, and, in *Harmonie universelle* (1636–1637), he discussed methods for imitating human speech with apparatus as well as strategies for teaching speech to people who were deaf. Yet, although Sauveur experienced hearing and speech difficulties himself, he appeared less interested in examining the physiological or philosophical dimensions of acoustics. Rather, Sauveur followed in Mersenne's footsteps in regard to scientific investigation of the phenomena of sounds. Sauveur classified temperaments of the musical scale and divided the octaves into smaller units. He identified and compared various methods of tuning and he introduced definitions of pitch standards. He was also the first to employ beats to determine frequency differences. For this extensive work, Raman (1973) has recognized Sauveur as the founder of the science of acoustics.

Sauveur's challenge of interpreting musical sounds through the use of "musical assistants" must have been analogous to the English chemist John Dalton's task of investigating color changes. Dalton was color-blind, a condition which he first observed in himself and described in a publication in 1794. Similarly, the color-blind German mineralogist Ferdinand Reich employed an assistant, Hieronymus Richter, who worked under his direction in examining metals spectroscopically. Through Richter's eyes, Reich discovered an indigo-colored line, and this provided evidence of a new chemical element which was subsequently named indium.

Charles de la Condamine (1701–1774), although admitted to the Academy as a chemist, is better known for his contributions as a geophysicist and explorer. He made the first scientific exploration of the Amazon in 1743. Through the study of Roman remains he attempted to define the dimensions of linear measure and, during an expedition to Peru, he measured a meridional arc at the equator. Using a star as a reference to make observations of a plumb line both near and away from a mountain mass, he demonstrated that the mountain exerted an attraction on the plumb bob (Williams, 1982, p. 301). His friend and fellow scientist Comte de Tresson, author of *A Treatise on Electricity* (1749), once described de la Condamine as "very deaf and very importunate, the terror of the members of the Academie, from the vivacity with which he urged inquiries, which could only be satisfied by the inconvenient medium of his hearing trumpet" (*VR*, *27*[7], 1925, p. 343). The terror of de la Condamine may not have been his bold inquiries alone. De Tresson's choice of words to describe the communication process appears to indicate that the inconvenience in satisfying the inquiries was, to a lesser degree, de la Condamine's. As de Tresson explained, the deaf novelist Alain René le Sage had prepared him "to sustain the petulant activity of the hearing trumpet of my deaf and illustrious associate and friend, Mons. de la Condamine" (p. 343). Charles de la Condamine died during an operation for paralysis. True scientist that he was, he had set out to observe the details of the operation with intent to publish.

Jean-Jacques Rousseau (1712–1778), the son of a watchmaker in Geneva, had invented a musical notation which brought him into friendship with Voltaire and Buffon. His companionship with Diderot encouraged him to join the struggle against the social injustice of his day. His powerful call for political reform and democracy emphasized the balance between individual liberty and social organization. In *Social Contract* (1762), Rousseau expounded Locke's assumptions of natural law. His views on the importance of equality and justice later formed, in part, the philosophical basis of the American Revolution. In *Émile*, he put forth a doctrine of education with a then-startling emphasis on children's rights and a naturalistic view of human development; and this brought him under persecution by his opponents. Cautious in their support of him were his own friends, including de la Condamine.

Rousseau's emotional description of the onset of his deafness and tinnitus (head noises) at the age of twenty-four can be found in Book VI of *Confessions*, an "internal tumult" so violent that it "has injured my auricular organs, and rendered me, from that time, not entirely deaf, but hard of hearing." Biographer Gustaf Vogt (1939) believed that Rousseau's deafness contributed to his shyness and explained his avoidance of large audiences: "He once escaped by actual flight an audience with the king. He preferred to bury himself in the solitude of nature where he learned to express his thoughts in writing with a courage unheard of in those times" (p. 29).

In the years following Rousseau's death, his *Letters on the Elements of Botany to a Lady* went through eight editions and had widespread influence on the movement toward improving educational conditions for women in science.

The Academy's Role in the Education of Deaf Children

Only a few decades after the death of Sauveur and Amontons, members of the French Academy of Sciences began to examine the potential of deaf youth to learn. Through these efforts, the instruction of deaf pupils gained respect as a profession. Among those instructors invited to the Academy for demonstrations in the 1740s was Jacobo Rodriguez Pereire, who had successfully taught several deaf youth, including Aaron Beaumarin and Azy d'Etavigny. D'Etavigny, the son of the director of leases of the town of La Rochelle, accomplished enough to stir the Academy. When he was about thirteen years old he was placed under the charge of Father Cazeaux at Beaumont. After both Cazeaux and Étienne de Fay did not succeed in educating D'Etavigny, Pereire stepped in and, using pronunciation, signs, fingerspelling, and speechreading, impressed the distinguished Comte de Buffon, who had joined others in the Royal Academy of Sciences in examining D'Etavigny in 1749.

Next came the deaf pupil Saboureux de Fontenay, who was subjected to extensive questioning on January 21, 1751, by Buffon, the physician Antoine Ferrein, and the physicist Jean-Jacques Mairan. Ferrein's scientific interests included the modification of Dodart's theory of phonation. Marain, who also had an interest in acoustics, was coincidentally confirming around this time the observations made earlier by the deaf physicist Guillaume Amontons that a liquid cools while evaporating. When he was examined, the deaf pupil de Fontenay was thirteen years old, the son of a colonel of the light calvary of the Royal Guards. He had an impressive mind and it is no surprise that upon observing him, the eminent members of the Academy concluded that deaf persons could indeed be educated.

De Fontenay's enthusiasm in science is shown in the comments of Charles Michel Abbé de l'Epée who was responsible for the great gains in the education of deaf people in eighteenth-century France. Epée once remarked that Saboureux de Fontenay maintained "by means of writing, regular disputations not only on the different subjects of common conversation, but even upon sciences of which the generality of mankind are ignorant. I have seen him dispute for half an hour with another on the generation of plants and the culture of mushrooms" (Braddock, 1975, p. 44). De Fontenay himself had commented that his instructor Pereire "enjoyed taking me to see experiments in physics, collections of scientific curiosities and so on" (Lane, 1984, p. 81). His long-standing interests in science appeared to complement his contributions in the philosophical study of languages. In 1780, he authored a memoir on meteorology which was presented to the Royal Academy of Sciences. This may be the first evidence found in historical records of a congenitally deaf person receiving an education adequate for the pursuit of work in physics.

De Fontenay associated with some of the greatest thinkers of his time. Urged by the Duke de Chalnes, he published in the *Journal de Verdun* (1765) a description of how he had come to master language. "You will marvel with me," he explained, "that a person who cannot speak should write thus on languages in general, and should desire to trace them back to their origin. It is a case exactly similar to that of the blind [Professor Nicholas] Saunderson, who wrote on colors and on heavenly bodies" (*AAD*, *23*[1], 1878, p. 37). Through the eyes of Saunderson, de Fontenay's friend Denis Diderot had earlier argued against using design in nature as proof of God's existence.

In 1770 de Fontenay traveled to Sweden to study Arabic under Professor Bjoernstaehl at the University of Uppsala. He had mastered the grammar of Hebrew, Syriac, and various European languages, and it is remarkable that he had accomplished this without formal schooling. He brought with him a letter of support written by the French naturalist Duchesne, an associate of the great botanist Carolus Linnaeus and an author of a treatise on botany himself. In this letter of recommendation to Professor Bjoernstaehl, Duchesne explained that although de Fontenay received instruction from Pereire, "he owes only to his own efforts the immense fund of information of all kinds which he has acquired" (*AAD*, *23*[1], 1878, p. 38). His congenitally deaf student so impressed him that Bjoernstaehl, with the aid of a translator, spread word of de Fontenay's success while traveling through Germany. Bjoernstaehl published his account in 1773 in Stockholm, which was translated into German in 1777 and published in Stralsund.

We are left to wonder if Charles de la Condamine and Jean-Jacques Rousseau involved themselves in the demonstrations of deaf pupils in part because of their personal experiences with deafness. La Condamine arranged for Pereire's pupils, Marie Marois, Le Rat de Magnitot, and Mlle. de la Voute, to be presented to the king of Sweden. Rousseau worked closely with Buffon to monitor the progress of Marois.

Without doubt, more than one revolution was brewing in France. In the same year (1751) that de Fontenay was examined at the Academy, Denis Diderot not only published the first volume of his thirty-five volume *Encyclopédie*, for which he is best known, but also authored a "Lettre sur les sourds et muets," which described the social barriers experienced by persons with disabilities in the eighteenth century. As Mettler (1947) has explained, Diderot's work "played a definite role in abolishing the old idea that the congenitally blind or the deaf-mute [sic] is mentally deficient, and pointed out that man's ideas are the result of experience gained by means of the five senses" (p. 503). Slowly, and under the influence of the great Romantic philosophers Rousseau, Diderot, and others, the isolated attempts to teach deaf youth gave way to observable gains and by the 1760s France, under the guidance of the Abbé de l'Epée, had established the world's first government-sponsored school for deaf children.

As is true today, one cannot distinguish deaf and hearing scientists of the Enlightenment by their writings. The "invisibility" of deaf authors in the eighteenth century is well illustrated in the argument that occurred in 1779 when Pierre Desloges, a glue worker and bookbinder, published a letter advocating sign language in *Mercure de France* (addressed to Antoine-Nicholas Condorcet, the perpetual secretary of the Academy of Sciences), and published a book on this topic, *Observations of a Deaf Man on an Elementary Course of Education of Deaf and Mute Persons*. When Desloges' book was claimed to be the first by a deaf author, Saboureux de Fontenay wrote to Desloges to protest that his own autobiography, published in 1765, had preceded Desloges' work. Yet, unknown to both of them were numerous books by the Swiss naturalist Charles Bonnet. And Guillaume Amontons' only book, *Remarques et expériences physiques sur la construction d'une nouvelle clepsydre*, was published even earlier in Paris, in 1695.

ENGLAND AND THE ROYAL SOCIETY

The deaf experience in the French Academy was unique, both with respect to the examination of deaf pupils and the coincidental deafness of several of its members. Across the English Channel, the Royal Society's long-standing interest in the medical and educational aspects of deafness was punctuated with the seventeenth-century work of such well-known hearing personalities as John Wallis, the brilliant mathematician and Lucasian professor of Oxford who helped to found the Society, the great chemist and physicist Robert Boyle, author of *The Sceptical Chymist* (1661), who encouraged the systematic study of the elements, and Christopher Wren, the architect who helped rebuild London after the great fire. The rich history of their involvement can be found in the *Philosophical Transactions*. As with the French, there were the numerous demonstrations associated with hearing and deafness, such as the diarist Samuel Pepys's inspection of "a great glass bottle broke at the bottom" which, when he put the neck to his ear, allowed him to "plainly heare the dancing of the oares of the boats in the Thames"; or Richard Waller's description of a brother and sister about fifty years old who "had not the least sense of hearing, yet both of them understood by the motion of the lips only, what was said to them." It was not until the eighteenth century, however, that we find several corresponding

Charles Bonnet, Swiss Naturalist and Philosopher

members of the Royal Society who were deaf. Charles de la Condamine was elected in 1749, but his work was primarily addressed to the Royal Academy in his own country, France. The Swiss naturalist Charles Bonnet (1720–1793), a corresponding member of the French Academy, was also elected a fellow of the Royal Society in 1743. Deaf since seven years of age, Bonnet wrote in his *Memoirs* of his profound experience with hearing loss. "It was my deafness, which had already begun to manifest itself," he explained, "that frequently made me the object of scorn in the lessons and lectures and exposed me without stop to ridicule" (Savioz, 1948, p. 41). Bonnet was taken out of school by his parents and assigned a private tutor, and it was then that his interest in natural science began to develop through his reading at home.

At the end of the seventeenth century, Anton van Leeuwenhoek had searched without success for evidence of eggs among insects, noting that aphids brought forth their young alive. René Réaumur, extending Leeuwenhoek's work with both winged and wingless aphids, published his *Histoire des insectes* in 1734. This book was eagerly read by Bonnet, and the work of Réaumur and that of Abbé Pluche, author of *Spectacle de la nature*, had a great influence on him. After several years of studying, Bonnet began a correspondence with Réaumur and asked the naturalist to suggest a subject for investigation. Bonnet's investigations of a female spindle-tree aphid led to his discovery of reproduction without fertilization, a process biologists call parthenogenesis, and his subsequent publication of *Traite d'insectologie* in 1745 earned him recognition as one of the first experimental entomologists.

Yet, unlike the practical observations of de la Condamine and despite his experience as a deaf child, the extent of Bonnet's interest in the education of deaf pupils appears to have been limited to his later philosophical writings, such as in his discussion of the use of signs in *Essai analytique sur les facultés de l'âme* or in his essays on language and rational thought in *Essai de psychologie* and *Contemplation*. Also almost completely blind since his youth, he employed assistants to conduct some of his research, publishing many treatises on botany and his early views on evolution. He was a pioneer in the study of photosynthesis and made suggestions to Lazzaro Spallanzani on the artificial insemination of a dog.

The earlier observations of several hearing members of the Royal Society merit mention. These men studied the abilities of deaf pupils to communicate and to learn, and their essays on instruction in speech, the use of gestures, sign language, and manual alphabets indicated that all of these methods were recognized, to various degrees, as viable educational tools. Much earlier, Francis Bacon had discussed how gestures could be used to express ideas and had examined the use of signs and their referents. John Wilkins, one of the most important intellectual statesmen of science in his time, followed in Bacon's footsteps. He was the brother-in-law of Oliver Cromwell and close friend of the diarist John Evelyn. Wilkins did much to formulate the scientific aims which had great influence on his fellow scientists' perspectives of their roles in social life. His writings included a pamphlet describing the potential of sign language for teaching deaf children. The noted English architect Christopher Wren, designer of St. Paul's Cathedral, was the most brilliant pupil of Wilkins. Perhaps for his use, Wren's sister is said to have invented a manual alphabet for people who were deaf. The involvement of Royal Society members took an unpleasant turn, however, with the bitter dispute between John Wallis and William Holder over their individual claims as the first teacher of deaf students in England, a controversy that was for some time the focal point of their publications in the *Philosophical Transactions.*[8] When Wallis reported on his success with two young deaf men, Daniel Whalley and Alexander Popham—one of whom had been instructed earlier by Holder—he made no mention of Holder's work. Holder's later attack of Wallis in the *Philosophical Transactions* was followed by an equally hostile reply by Wallis entitled "A Defence of the Royal Society, and the Philosophical Transactions, Particularly Those of July 1670, in Answer to the Cavils of Dr. William Holder." The dispute was noted by such Englishmen as the diarist Samuel Pepys and the Bishop of Salisbury, Gilbert Burnett. Wallis, despite his prolific success as a Lucasian professor of mathematics and associate of Newton, holds the dubious distinction of having had the first public quarrel in the education of deaf persons, and as Hodgson (1953, p. 101) has written, "This petty tendency was unfortunately to remain a feature of the work."

As time went on, scientists in the Royal Society continued to take a leadership role in the early efforts to educate deaf children. Henry Baker (1698–1774), a well-known hearing microscopist often noted for popularization of its use, found success in instructing deaf persons in speech. The Huguenot refugee and fellow of the Royal Society Louis Dutens described Baker's efforts with an eighteen year-old deaf woman in *Memoirs of a Traveller*: "I applied to a professional man, named Baker, who by a method of his own had taught Lady Inchiquin and her sister, and some other pupils. . . . I saw some of his scholars; and was astonished at the facility with which they understood what I said, by observing the motion of the lips: they also answered me, but their voices wanted modulation" (*Christian Observer*, *8*, 1809). Baker's methods were not his own, but he guarded them nonetheless with utmost care. The writer Samuel Johnson encouraged him to publish them, but Baker had made a profession of teaching, and protecting his methods allowed him to continue to charge high fees for his services, a practice which helped him earn a considerable fortune. In fact, Baker charged his pupils each 100 pounds as security not to reveal the methods he used to teach them. The great pains which he took to keep his methods secret is revealed in the collection of fifty legal agreements housed in the Manchester University Library Special Collection, dated 1735 to 1758, all concerned with promises of secrecy and duly witnessed.[9] When Baker's notes were later located in the archives of the Royal Microscopical Society, it was learned that his methods were quite similar to those promoted by John Wallis. But this is not surprising. In 1729 Baker married the novelist Daniel Defoe's daughter. One of Defoe's novels, *The History of the Life and Adventures of Mr. Duncan Campbell* (1720), was a story about a deaf English gentleman who was instructed by an acquaintance of Wallis, and the book was used to promote Wallis's reputation. Wallis was Daniel Defoe's brother-in-law.

THE EDUCATION OF JOHN GOODRICKE

Until this point, only fragmented accounts of deaf people in science have been presented. This is particularly true for those who were born deaf. Much more is known about their instructors. An examination of the education of John Goodricke (1764–1786) of York, England, however, provides a powerful lesson in the effects a quality education can have on the self-esteem and subsequent success of a congenitally deaf child. It begins nearly a century after John Wallis provided his account of his work with deaf pupils to Robert Boyle, then secretary of the Royal Society. Wallis's writings fell into the hands of a merchant of Leith, Charles Shirreff, the father of a deaf boy. What eventually ensued was probably the first practical result to come out of Wallis's publication. Shirreff, determined to find someone who would establish a school in which his son might be similarly educated, encouraged Thomas Braidwood, who opened an academy in Edinburgh in 1760. Thirteen years later, in 1773, the writer Samuel Johnson attested to Braidwood's success. After visiting the Braidwood Academy, Johnson provided an account in *Journey to the Western Islands*, first published in 1775. Johnson described how he had found a group of deaf children waiting for their master, "whom they are said to receive, at his entrance, with smiling countenances and sparkling eyes—delighted with the hope of new ideas." Johnson was himself rather deaf since childhood from scrofula, and this experience at the Braidwood Academy must have held special interest for him. During this visit, he challenged a young deaf girl by writing on her slate a multiplication problem consisting of three figures multiplied by two figures. "She looked upon it, and, quivering her fingers in a manner which I thought very pretty," Johnson explained, "but of which I knew not whether it was art or play, multiplied the sum regularly in two lines, observing the decimal place." Comparing the experiences of two isolated groups, Johnson wrote that "after having seen the deaf taught arithmetic, who would be afraid to cultivate the Hebrides?"[10]

This story about a deaf child learning arithmetic is not unusual. For a century or more there had been accounts of the mathematical abilities of young deaf children, but until the latter quarter of the eighteenth century, no reports can be found of any of these children bringing such skills to fruition in a scientific profession. Among the children attending the Braidwood Academy when Johnson visited on that day in 1773 was a nine-year-old boy, born deaf, whose work would lay the foundation for the study of binary stars. As records show, John Goodricke entered the Braidwood Academy a year before Johnson's visit. Applying his mathematical talents and observational powers, Goodricke's study of the star system Algol in the constellation Perseus earned him, by the age of twenty-two, the Godfrey Copley Medal from the Royal Society.

More is known about John Goodricke than any other congenitally deaf person of his time, but there are scant records of his early life. John's father made sure he was baptized in an Anglican church two days after his birth. His hearing loss was so profound that "he was not aware of voices at all and as a consequence was unable to learn to speak" (Richardson, 1967, p. 85). As Richardson wrote, "In the eighteenth century, a boy born with such a crushing handicap could hope to survive by performing only the most menial type of labor" (p. 86). But this was not to be the case for Goodricke, who may have been a genius. Living in a period in which the quality of education for deaf children was far from adequate, his natural ability and the wealth of his parents combined to help him obtain an education which few would have expected possible of a congenitally deaf youth. Even a century after Goodricke's death, the majority of boys in the schools for deaf pupils were being prepared for such trades as shoemaking and tailoring, the girls for needlework and other handicrafts. A similar state of affairs was found in other countries. In Germany in 1886, for example, the most common occupations for deaf people were clothiers, cleaners, farmers, and graziers. Half as many were working as "all sorts of personal servants."[11] Only six deaf people in this report were listed in teaching positions, and four others in art, literature, and journalism. There were no deaf people identified in any field of science. The same sad situation is found in the statistics provided by officials of other European countries and the United States.

During my visiting lectureship at the University of Leeds in 1988, I searched for records of Goodricke's education. I was not surprised to learn that he was a member of a wealthy family. His father, Henry Goodricke, served in a diplomatic post in Holland. A literary man, Henry published a Latin work on jurisprudence in Groningen when John was two years old. Henry's father, Sir John Goodricke, was for thirteen years envoy extraordinary for Britain at the court of Stockholm, and later a privy councillor to George III. In Carolyn Gilman's account of the astronomer John Goodricke, she wrote, "The Goodrickes were an enlightened family. They were solid Yorkshire gentry, graced with a hereditary baronetcy, a country manor house, and a fine stable of horses" (1978, p. 400).

Goodricke's achievement in the Braidwood Academy must have been at the very least satisfactory; in 1778 he was able to enter the Warrington Academy, a well-known nonconformist educational institution in the north of England. Warrington provided no special provisions for students with hearing loss. During that year, Goodricke was one of twenty-seven students, and he likely received a well-rounded and full course of study by excellent teachers, including lessons in Greek and Latin. Since the Braidwoods charged high fees and Warrington Academy was also expensive, such an education which contributed so much to Goodricke's success was accessible only to the children of the wealthiest of families.

Working at first with a crude telescope made from spectacle lenses, Goodricke followed the suggestion of his hearing friend Edward Pigott and began a systematic study of Algol, the "demon star," in Perseus, and he surprised prominent astronomers with his explanations of possible reasons for its changing brightness. Through letters written to Royal Society members, Goodricke generated much interest in the notion of binary stars, and published in the *Philosophical Transactions* his observations on Algol, Delta in Cepheus, and other stars. He also corresponded with William Herschel, the discoverer of Uranus, about his astronomical interests. "I have read your curious paper on the construction of the Heavens with great pleasure," he wrote to Herschel. "It seems as if all the riches of the heavens are now opened to us by means of your

John Goodricke, British Astronomer

large telescopes and I heartily wish you success in the farther pursuit of the subject" (C. A. Lubbock, 1933, pp. 195–196).

In regard to his own measurements of the brightness of Algol, Goodricke wrote a letter to Anthony Shepherd, Plumian Professor at Cambridge, which was subsequently read to the Royal Society on May 12, 1783, and published in the *Philosophical Transactions* as "A series of observations on, and a discovery of, the period of variation of light of the bright star in the head of Medusa, called Algol." He noted that Algol appeared to have a companion, and that the system eclipsed itself at regular intervals:

> If it were not perhaps too early to hazard even a conjecture on the cause of this variation, I should imagine it could hardly be accounted for otherwise than either by the interposition of a large body revolving around Algol, or some kind of motion of its own, whereby part of its body, covered with spots or such like matter, is periodically turned towards the earth. But the intention of this paper is to communicate facts not conjectures; and I flatter myself that the former are remarkable enough to deserve the attention and farther investigation of astronomers. (Goodricke, 1783, p. 482)

Goodricke's education prepared him well to perform observations and to propose possible explanations of the phenomena. However, Storm Dunlop (1981), Fellow of the Royal Astronomical Society, conjectured in 1961 that we might wonder if Braidwood had actually succeeded in teaching Goodricke to speak, after reading a letter from Goodricke's close friend and fellow astronomer Edward Pigott, in which Pigott commented that "there is not a soul here to converse with" on astronomy. Nevertheless, nearly a century before the German astronomer Hermann Vogel used spectrographic analysis to confirm that Algol was a binary star, young John Goodricke broke through the silence of his world to alert the scientific community to the conception of two-world systems in harmonious motion.

The Braidwood Academy is described in nearly every history of the education of deaf people, but John Goodricke is a name few educators of deaf children will recognize. Even in York, England, where I lived for three months, Goodricke's name brought no recognition among the members of the local Deaf community or educators in the schools. Yet, as one of the first persons deafened in infancy to contribute significantly to a field of science, John Goodricke was himself a new star, and a fitting tribute to his work has been long overdue.

DEAF NATURALISTS IN EIGHTEENTH-CENTURY ENGLAND

The most noteworthy of the British naturalists with deafness during the Enlightenment was the Reverend Gilbert White (1720–1793) whose classic work *The Natural History and Antiquities of Selborne* was published in 1789. Walter Johnson (1928) has written that it is uncertain precisely when White's deafness began: "Probably both deafness and failing sight were evil legacies from the attack of smallpox which he endured when a young man of five and twenty" (p. 38). After White's death, an ear trumpet was found among his effects (Holt-

White, 1901, p. 39). White was educated at Basingstoke by Thomas Warton, who, incidentally, had two deaf brothers. In a letter dated September 13, 1774, fifteen years before he published *Natural History*, White described his fluctuating hearing loss:

> Frequent returns of deafness incommode me sadly and half disqualify me for a naturalist; for when those fits are upon me I lose all the pleasing notices and little intimations arising from rural sounds; and May is to me as silent and mute with respect to the notes of birds, etc., as August. My eyesight is, thank God, quick and good, but with respect to the other sense, I am, at times, disabled; and wisdom at one entrance, the ear, is quite shut out. (Roe, 1917, p. 364)

The gloom in White's writing is understandable, and reinforces what I have written earlier about Charles Butler's late deafness. White's *Natural History* was based on correspondence with the zoologist Thomas Pennant and other notables and his diary of daily notes on natural phenomena, especially birds, which he observed in his garden at his large house in the center of Selborne village and during his walks into the countryside. Much of the material needs no correction today. He recorded notes on the many species of wildflowers. He was the first to recognize Britain's smallest mammal, the harvest mouse, and had an intense interest in bird life, recording his observations of anatomy, plumage, and habitats. While Linnaeus had first described the three British leaf warblers (*Phylloscopi*) and the similarity in their plumage, it was White who emphasized that their songs were different. White's observation that the domestic pigeon had stemmed from the blue rock pigeon, and not from the wood pigeon or the stock dove as had been conjectured earlier, was elaborated on by Darwin as he prepared his theory of evolution (Groves, 1976, p. 299). In his boyhood, Darwin read White's *Natural History* and took up the study of the habits of birds as a consequence. Many years later he "made a pilgrimage to the shrine of Gilbert White at Selborne" (Darwin, 1887, p. 35).

Another British naturalist, Cyril Carr of Sheffield, lost his hearing through fever when three years old. As a member of the Microscopical Society, he lectured on such topics as the "Biology of Starch" and the "Physiology of Blood," and presented lectures to the Deaf community as well (Roe, 1917, p. 351). Brief reports in the literature of the Deaf community such as this one about Cyril Carr have focused less on the scientific work and more on clever solutions to the communication barrier which deaf persons have often faced when in circles of hearing people. They provide us with little information about these individuals as scientists. Another illustration of this was the description of a congenitally deaf man whose last name was Mackenzie, "particularly distinguished by his enthusiasm and attainments in natural history." Mackenzie's attainments were not explained. Rather, the emphasis of this report was on how he had encouraged fellow naturalists and other visitors at his house in Stornoway, in the Isle of Lewis, to become adept at fingerspelling in the 1780s. The Reverend John Buchanan summarized the communication as follows, "After seeing a few letters spelt on [a visitor's] fingers, [Mackenzie] immediately supplies the rest and saves them the trouble of going through the whole. Those who have the honour of visiting at his house are at pains to touch their fingers cleverly . . . in order to make themselves understood when in company with him" (Roe, 1917, p. 356).

DEAF MEMBERS OF OTHER SOCIETIES

Scattered reports of other persons with deafness can be found in the literature, including an army surgeon named Alexander Small who was a member of the Society of Arts, Thomas Lawrence, the president of the Royal College of Physicians of London, and Aloys Weissenbach, an Austrian field army surgeon.[12]

One of the most important deaf contributors to a field of science in the eighteenth century was Anders Gustaf Ekeberg (1767–1813) of Sweden. Uppsala was a center for chemical research in Ekeberg's time. By the 1770s, Torbern Olof Bergman was sponsoring Karl Wilhelm Scheele's investigations, which led to many discoveries in analytic chemistry. In 1789 Antoine-Laurent Lavoisier published his great work *Elementary Treatise on Chemistry*, renewing Robert Boyle's suggestion that the elements were material substances which could be broken down and examined under controlled experimentation, and the principles for assigning names to chemical substances on the basis of their composition had been established. The early deaths of Bergman and Scheele left Sweden with no outstanding chemists to enrich the science, and it was at this time that Ekeberg labored to strengthen Lavoisier's pioneering work. Ekeberg, progressively deafened since the age of eleven, and having lost the use of one eye in a chemical explosion, was not deterred from stepping in and filling this honor. In 1803, he discovered the element tantalum, and he was elected a member of both the Royal Academy of Sweden and the Academy of Sciences in Uppsala for his chemical work. Ekeberg enjoyed public-interest science, lecturing on such topics as combustion and the benefits of chemical research for medicine. He was a talented poet as well, and in 1801 he wrote a poem in invisible ink which he presented to the king of Sweden, who warmed the paper and read the deaf scientist's words describing his desire for a world of peace. Ekeberg was also an excellent teacher. One of his students, Jacob Berzelius, went on to discover selenium, silicon, thorium, and several other elements.

Ekeberg had isolated tantalum in a mineral and he conducted extensive analyses of its properties. His discovery, however, was challenged eight years later by William Wollaston, who mistakenly declared that the element columbium identified earlier by Charles Hatchett was identical to tantalum. In 1809, Ekeberg sent Thomas Thomson samples of tantalite but the ship carrying these samples sunk in the Baltic. After Ekeberg's death, Berzelius corresponded with Thomson and defended Ekeberg's discovery. Jean Marignac later used spectroscopic analysis to confirm that they were distinct metals, and columbium was eventually renamed niobium (Niobe being the daughter of the mythical Tantalus).

Anders Gustaf Ekeberg, Swedish Chemist

AMERICA'S EARLY SOCIETIES AND THE EDUCATION OF DEAF CHILDREN

Frederick E. Brasch (1931), chief of the Smithsonian Division, Library of Congress, wrote that there is nothing that has contributed more to the progress of science since the Renaissance of learning in Western Europe than the growth of scientific societies and academies. In his article, "The Royal Society and Its Influence upon Scientific Thought in the American Colonies," Brasch described the contributions of American colonists who had been honored with membership in the Royal Society, and how the Society's interests were reflected in the work of the colonists as well. Names of Royal Society members that stand out in both the history of science and the history of the education of deaf pupils have been mentioned previously, most notably Wallis, Holder, Boyle, Wilkins, and Wren. In France, the early work of Buffon, de la Condamine, and Rousseau in the Academy of Sciences brought attention to the instruction of Pereire and others and paved the way for the great advances of Epée. The early American colonists were certainly cognizant of the observations of the Royal Society members on deafness and deaf people. Cotton Mather, for example, an American member of the Royal Society and an influential member of the American Philosophical Society, was a close correspondent with Richard Waller, the Royal Society's secretary, who reported his observations of profoundly deaf people who could speechread. But to translate such observations to useful endeavor in the colonies was no easy task. Mather himself was persecuted by the Puritan thinkers, for example, when, having secured information from the Royal Society's *Philosophical Transactions* on fighting smallpox, he pioneered in preparing inoculation in New England. Similarly, there was doubtless a need to learn more about teaching deaf children from the Europeans. In 1680, George Dalgarno of England made the linguistically perceptive comment about sign language that there might be "successful addresses made to a [deaf] child even in its cradle" if parents had "but as nimble a hand as usually they have a tongue." For this observation about the need to communicate with deaf infants, Dalgarno may be considered far ahead of his time. But, in the colonies of the New World, there was marked contrast in views. Only a year earlier, in 1679, Philip Nelson had endeavored to "cure" a deaf child (perhaps this meant he had tried to teach him speech) and faced the wrath of the people of Rowley, Massachusetts. Deafness, in the minds of many of the American colonists, was predestined; and attempts to teach deaf children were met with accusations of sorcery and witchcraft.

During these early years, members of the American Philosophical Society discussed issues in natural philosophy, history, and politics, and conducted studies in botany, medicine, chemistry and other scientific areas, their first major investigation being the transit of Venus in 1769. But, there was no concerted effort of the Society to study the educational needs of deaf pupils and, as the Enlightenment came to a close, there remained no formal opportunities for deaf children to learn in the colonies. Hence, a deaf nephew of President James Monroe was sent to Paris for his education, and Colonel Thomas Bolling was forced to send his deaf son to the Braidwood Academy in Edinburgh. A few years later, he sent two more of his deaf children there. The education of these children, as well as that of Francis Green's son (also at the Braidwood Academy), represents the earliest documented efforts to formally educate American children who were deaf. A century after the Salem witchcraft trials, in 1783, Green published *Vox Oculis Subjecta* ("voice made subject to the eyes"). The title of this report, the motto of the Braidwood Academy, reflected Green's appreciation for the school that had succeeded so well in instructing his son.

In 1784, another American observed the Abbé Sicard's classes in Paris and wrote a letter to Judge William Cranch in which he praised the work of Epée and his followers: "His success has been astonishing; he teaches the deaf and dumb [sic], not only to converse with each other by signs, but to read and write, and comprehend the most abstracted metaphysical ideas" (*AAD*, *8*[2], 1856, p. 248). The seventeen-year-old American youth who wrote this was John Quincy Adams, who would become president of the United States and a very active member in the National Academy of Sciences. As president, Adams was later generous in his support in establishing the first school for the deaf west of the Alleghenies (Kentucky).

The otherwise-enlightened Thomas Jefferson, however, a member of the American Academy of Arts and Sciences, patron of the Columbian Chemical Society of Philadelphia, and second president of the American Philosophical Society, would argue against a program for the education of deaf children when it was proposed for consideration in the plans for the University of Virginia. Jefferson, the university's founder, believed that it would be "mere charity" and that such efforts with deaf children might impede the work with hearing children (*AAD*, *42*[2], 1897, pp. 119–120). I cannot help but wonder if Jefferson might have felt differently had he known that his own great-grandson, Thomas Jefferson Trist, born in Monticello, Virginia, in 1818, would attend the New York Institution for the Deaf.

Coincidentally, one of Benjamin Franklin's lineal descendents, Catherine W. Bache, would one day also be a pupil in this school. Franklin was the first president of the American Philosophical Society and he was, of course, interested in nearly everything, including deafness. In *Vox Oculis Subjecta*, Francis Green mentioned Franklin as one "among the many who have attended the public examination and attested the progress of several pupils of the justly celebrated Mr. Braidwood of Edinburgh (who hath brought this very curious, important, and almost incredible art to a much greater degree of perfection than any former professor)" (p. 12).

Even though the American Philosophical Society honored such eminent foreign scientists as Buffon, Condorcet, and Herschel with membership, all with personal experiences with educated deaf people, and such prominent members as Franklin and Adams had firsthand observations, the role of the early American scientific societies in promoting the education of deaf pupils was not nearly as visible or as strong as that of their predecessors in England and France. The American Philosophical Society does hold the honor of having been the first scientific society in the colonies to publish a report on teaching deaf children. In 1793, William Thornton, head of the U. S. Patent Office, published "Cadmus, or a Treatise on the Elements of Written Language" with "Essay on the Mode of

Teaching the Deaf and Dumb, to Speak" in the *Transactions of the American Philosophical Society.*

The death of his own deaf son by drowning had not stopped Francis Green from pursuing this cause. In 1803, Green published in a Boston newspaper a request to the clergy in Massachusetts to obtain information on the number of deaf children residing in the state. It was his intention to determine whether the number warranted the establishment of a school. In the following year, the Reverend John Stanford found several deaf children in an almshouse in New York City and attempted to teach them. A decade later, another clergyman, Thomas Hopkins Gallaudet, attempted to teach the deaf daughter of his neighbor, a New England physician, Mason Fitch Cogswell, who gathered enough financial support to send Gallaudet to Europe to study the methods employed in the well-known schools begun by Thomas Braidwood and Charles Michel Abbé de l'Epée. The Braidwoods refused to share their methods with Gallaudet, but while in London, Gallaudet was fortunate to have the opportunity to attend a lecture by Abbé Sicard, which included demonstrations by two successful deaf pupils, Jean Massieu and Laurent Clerc. Gallaudet was already familiar with Sicard's work, having read a publication procured by Dr. Cogswell, and he had learned some signs and fingerspelling while attempting to teach Alice Cogswell. Sicard invited the American to the National Institution for Deaf-Mutes in Paris, where Gallaudet subsequently spent several months. Here, Gallaudet was able to convince Laurent Clerc, a thirty-year-old assistant teacher, to accompany him to Hartford, Connecticut. Back in America, funds were obtained and the Connecticut Asylum for the Deaf and Dumb (now named the American School for the Deaf) was established in 1817, with Gallaudet as its director, Laurent Clerc as the first deaf teacher of deaf students in America, and Alice Cogswell as one of the first seven pupils.

Gallaudet and Clerc's names are best known for their establishment of the first permanent school for deaf pupils in America. Nevertheless, a number of scientific society members stand out in this history. Samuel Latham Mitchill, the New York surgeon general and a prominent member of the Columbian Chemical Society of Philadelphia, the New York Society for the Promotion of Agriculture, Arts and Manufacture, and the short-lived American Mineralogical Society in New York City, was such an important personage. Prior to Gallaudet and Clerc's establishment of the school in Hartford, Mitchill had received a letter from another deaf man, Clerc's friend François Gard, who had offered himself as a teacher. Mitchill read Gard's letter, and that of Francis Green, and called a meeting at Tammany Hall. Within a short time, Mitchill found himself in the position of president of the New York Institution for the Instruction of the Deaf. His geological disciple and fellow physician, Samuel Akerly, was the Institution's new secretary. Neither man had much experience with deaf people and within a few years they had resigned. Akerly brought one aspect of his work with deaf pupils to the New York Lyceum of Natural History, presenting "Observations on the Language of Signs (among the Deaf and Dumb and the North-American Indians)" in 1823, and he published this paper in the *American Journal of Science* the following year. In 1818, Mitchill published *A Discourse*

Pronounced by Request of the Society for Instructing the Deaf and Dumb at the City Hall in the City of New York.[13]

Another notable American pioneer was Major Stephen H. Long, who, during the famous expedition to the Rocky Mountains in 1819 under the direction of John C. Calhoun, secretary of war, left Pittsburgh with the entomologist Thomas Say and the botanist and geologist Edwin James, and headed down the Ohio to its mouth, then up the Mississippi to St. Louis and then west. "The elucidation of a sign language [among Native Americans he observed]," wrote Long, "is particularly attractive to me, as connected with the interest of the institution in this place for the instruction of the deaf and dumb [sic], over which I have superintending care."[14] Thus, Long joined other prominent early scientists in the United States as an independent explorer in the education of deaf children.

This book has emphasized the theme that scientific societies, particularly those in France and England, have played important roles not only in bringing recognition to the education of deaf pupils as a profession, but also in providing a channel for deaf scientists to communicate. There was, of course, no dramatic increase in the number of deaf persons in the sciences, but those who were deaf, congenitally or adventitiously, were certainly aided by the rise of these societies. In much the same way as the Royal Academy had done with Saboureux de Fontenay and his work in meteorology a half century earlier, the presence of the Royal Society established a channel for communication which allowed John Goodricke to share his discoveries through the only means available to him—writing. Granted, their hearing contemporaries were similarly advantaged, but this does not lessen the importance for the deaf scientist of the Enlightenment.

2

The Nineteenth Century: Opportunities and Oppression

EDUCATION IN THE NEW WORLD

The original contract between Thomas Hopkins Gallaudet and the deaf teacher Laurent Clerc at the American School for the Deaf in Hartford, written in French, specified Clerc's general responsibilities in assisting Gallaudet in establishing the school and in maintaining instruction. In Article 11 of the contract, it was written that "Mr. Clerc shall endeavor to give his pupils a knowledge of grammar, language, arithmetic, the globe, geography, history . . ." (*AAD*, *24*[2], 1879, p. 116). Gallaudet agreed to take charge of all matters of religious teaching which might not be in accordance with Clerc's Roman Catholic beliefs. Science does not appear to have been taught at any level of importance during these early years at this institution, though Clerc was himself thoroughly schooled in science, as he once reflected in a presentation about his former teacher: "You see . . . what Mr. Sicard has achieved for his pupils. . . . The arts and sciences belong to the class of physical or intellectual objects; and the Deaf and Dumb, like men gifted with all their senses, may penetrate them according to the degree of intelligence which nature has granted them" (Waring, 1896, p. 9).

After Laurent Clerc, about twenty-five other deaf people played instrumental roles in the founding of educational institutions for deaf students, some becoming superintendents. Schools soon opened in Philadelphia (1820), Kentucky (1823), New York (1818, 1825), and Ohio (1827). By 1850 there were more than fifteen schools serving deaf pupils with nearly four out of every ten teachers in these schools deaf themselves. With the attendance of deaf male and female students at these residential schools and the increased use of sign language to teach them, the Deaf community in the United States was formed.

Many schools, particularly those in Connecticut and New York, also attracted fine young hearing instructors. At the American Asylum in Connecticut alone, at least twenty-nine Yale College graduates taught deaf children before 1890, several making their marks in both education and science.

Administering these schools along with the physicians Samuel Latham Mitchill and Samuel Akerly was the hearing scientist Oran W. Morris, who, while serving as principal at the New York Institution, conducted meteorological observations there as one of the most loyal of the 412 scientists in Joseph Henry's network established by the Smithsonian Institution. Later, when formal education for deaf children took a stronger hold, graduates of these schools, including the microscopist James H. Logan and the inventor Anson R. Spear, helped to establish additional institutions.

The quality of the science curricula in most of these schools is difficult to assess through analysis of the descriptions handed down to us. More information can be found on the adequacy of the curricula, however, during the years in which proposals for "High Schools" and "High Classes" for deaf pupils were presented at national conventions and published in the journals of the Deaf community. The New York Institution was a leader in the implementation of such a program. In 1851, two weeks before the death of Thomas Hopkins Gallaudet, W. W. Turner made a plea: "What [the deaf pupil] needs is a school expressly provided for him and for others in his circumstances, a High School for the Deaf. . . . [If it were] suitably endowed and judiciously managed, we might expect such a development of deaf-mute [sic] intellect as has not hitherto been witnessed in this or any other country" (*AAD*, *4*[1], p. 45). J. V. Nostrand then submitted for consideration of the convention of teachers of deaf children a proposal for selecting "from among the graduates of an institution those pupils whose proficiency in language and whose general characteristic for diligence and application to their studies made them candidates for the distinction, and offer them the privilege of a still further course of one or two years' instruction":

> Something must be done. Something that shall open to the mind of the deaf mute [sic] a wider range in the fields of knowledge than he has heretofore enjoyed; something to animate and excite him in the pursuit of knowledge, until he can take his place among the scholars and sages of the world. . . . The course of study for this class should embrace mental and moral philosophy, natural history, mathematics and natural philosophy, astronomy, history and English literature; in short . . . all the studies usually pursued in higher academies or even in colleges. (*AAD*, *3*[4], 1851, pp. 196–197)

In reading the biography of Thomas Jefferson's congenitally deaf great-grandson, Thomas Jefferson Trist, we learn that in this same year, 1851, he was knocked down by a locomotive and received serious head injuries. With immediate medical care, however, he was fortunate in being able to return to the New York Institution and, ranked second scholar, he graduated from the High Class in 1855. As Braddock (1975) explained, "The High Class of that decade was a sort of college for the deaf, requiring examinations in algebra (including binomial theorem and surds), first book of Euclid, natural philosophy, rhetoric and logic, and elementary chemistry" (p. 20). As discussed by Van Cleve and Crouch (1989), among those who supported a college for deaf students was the eminent deaf artist John Carlin, who in 1854 conducted a "careful and impartial investigation" at both the New York Institution and at the institution started by

Clerc and Gallaudet, the American School at Hartford, observing "prodigious strides" in various subjects, including astronomy, chemistry, and algebra (p. 77).

EARLY DEAF AMERICAN SCIENTISTS

Although Gallaudet College had yet to be established, there were deaf scientists scattered around the nation, a few born deaf and others adventitiously deafened. Several immigrants were among them. These men had begun to command authority in their respective fields. Frederick Augustus Porter Barnard (1809–1889) was a nineteenth-century Renaissance man with expertise and accomplishments as a mathematician, physicist, and chemist, as well as educator and poet. While he was at Yale College in 1824, his hearing began to fail. His brother and sister also experienced undoubtedly hereditary hearing losses, and his mother was "distressingly deaf" as well (Fulton, 1896, p. 54). While tutoring at Yale after studying under the notable chemistry professor Benjamin Silliman, he was frustrated that he could not hear his students across the room. As Fulton (1896) explained, "In the society of persons with whom he was not intimately acquainted he was much embarrassed, and he began to avoid society" (pp. 69–70). Fulton summarized:

> His own deafness came on at first so slightly that his friends could not believe it to be real. Barnard himself was only too willing to think he was the victim of a morbid fancy. The experience of the school-room soon proved that his deafness was neither imaginary nor temporary. It was real, and it steadily increased. He could still discharge his duties without serious inconvenience, but he allowed himself to cherish no illusions concerning the future. (pp. 54–55)

In 1831, encouraged by his classmate and friend David Ely Bartlett, Barnard began teaching deaf students at the school in Hartford established by Gallaudet and Clerc. In his National Academy of Sciences biographical memoir, Davenport (1939) described how deafness led to a change in Barnard's career plans: "Frederick Barnard, a scion of a family of professional men . . . originally trained for the law, was led, on account of a family defect in hearing, into teaching and administration. He maintained chemical and astronomical research as an avocation" (p. 263).

In 1832, Barnard accepted a position in the New York Institution for the Instruction of the Deaf and Dumb. As a teacher of deaf pupils, he entered upon the issue of the effects of deafness on language development, which led him to publish his *Analytic Grammar* in 1836. In this work he was closely associated with Leon Vaise and Harvey Prindle Peet. Vaise was an instructor from Paris who taught in the New York school for six years. Peet, another Yale graduate who had started teaching in the school in Hartford, was appointed the principal and superintendent of the New York Institution when the physicians Samuel Mitchill and Samuel Akerly resigned after a review by state authorities of the quality of instruction there. Peet went on to become a highly respected leader in the field of teaching deaf pupils. Barnard's friend Bartlett left Hartford and established a school of his own. Barnard took a half dozen deaf pupils with him to Virginia for demonstration and, as a result, induced the legislature to appropriate funds to open a new school at Staunton. He was also involved in the education of Julia Brace, a young girl both deaf and blind, whose success preceded that of the better-known Laura Bridgeman and Helen Keller.

Between 1837 and 1861, Barnard held positions at the University of Alabama and the University of Mississippi, and he served as president of Columbia College for twenty-five years (1864–1889). In his office was a large mechanical contraption he had built as an "amplifier" in which his colleagues would speak loudly, although it did not seem to help him all that much. Barnard published widely on such scientific subjects as the expenditure of heat in engines, pendulum motion, voltaic circuits, optics, hydraulics, gunpowder, the metric system, chemistry of metals, meteorology, and astronomy.

The private school founded by Bartlett was short-lived, but it may claim some credit for preparing a young boy who studied there to become the first deaf American to earn a Ph.D. While at the New York Institution, Bartlett had become dissatisfied with the fact that deaf children were unable to enroll in a school before the age of seven for additional help. He opened an early version of a "mainstream" program in 1852, integrating hearing and deaf students together and using both sign language and communication through speech and speechreading to optimize learning. Gideon E. Moore (1842–1895) attended the Bartlett School with his deaf brother, Henry Humphrey Moore, who was two years younger. Gideon preferred to use fingerspelling whenever possible. Henry later attended the Pennsylvania Institution for the Deaf in Philadelphia, but Gideon does not seem to have attended any other school until he entered Yale College, which graduated him with honors in chemistry in 1861. Moore lived by a motto he had etched with a diamond ring on one of the dormitory windows of Yale College. For many years after he graduated, this reminder of the lesson in fortitude he had provided his peers and professors was visible: "Perseverantia omnia vincit"—perseverance conquers all (Braddock, 1975).

In addition to chemistry, Moore studied German and after graduating from Yale he traveled to the University of Heidelberg. There he studied under the eminent scientists Gustav Kirchhoff and Robert Wilhelm Bunsen.[1] The physical laboratory at the University of Heidelberg was in the mid-nineteenth century a modest one, located in a small room of the "Riesengebäude," a house which was then 150 years old. In relation to Moore's unusual accomplishment, a reporter in *The Nation* stated that the deaf chemist had "learned his German by reading the lips of his teachers, and acquired his knowledge of the lectures of Professors Bunsen and Kirchhoff in the same manner." Moore earned his Ph.D. *summa cum laude* from the University of Heidelberg in 1869, and according to this reporter, he was the first American to be so honored. At the time, it was one of the highest scholastic degrees that could be obtained anywhere.

Gideon E. Moore published in both American and German science journals on his discoveries of minerals and the subsequent analyses of their properties. One new mineral he discovered was named "Brushite" in honor of George J. Brush, who had been his professor at Yale. Following the Civil War, Moore joined a team of hearing scientists from the National Academy of Sciences (NAS), which was investigating the production of sugar from sorghum.

The fact that Moore, the first deaf American to earn a doctoral degree in science, accomplished this within a decade of the first African-American (Edward Alexander Bouchet, Yale University, 1876) and the first American woman (Helen Magill, Boston University, 1877), leads one to continue such comparisons with African-American and women scientists as marginalized groups; but this is made difficult by the fact that it was not until recent decades that doctoral degrees have been earned by deaf persons in numbers that are significant. In 1974, for example, the *Gallaudet Almanac* listed only a half dozen deaf persons with earned doctorates in science-related fields. Of the 182 doctoral degrees held by deaf persons in the sciences which I have identified in my research for this book, more than 60 percent were earned by men and women either born deaf or deafened in their first five years.[2]

The documented experience of one of Moore's contemporaries, Robert J. Farquharson (1824–1884), illustrates the frustrations faced by deaf scientists communicating in group situations. Around the same time Moore was working with the NAS, Farquharson was presenting a paper at a meeting of the American Association for the Advancement of Science in Detroit. The anecdote which first attracted my attention to Farquharson's work (from the *New York Tribune*) follows:

> It often happens, as a matter of course, that the scientific gentlemen who read their communications are wanting in eloquence. More frequently still does it happen that after reading their first few sentences the demand of "Louder!" comes from their hearers. It generally has the needed effect. But when Prof. R. J. Farquharson began to read his really interesting paper on Recent Mound Explorations at Davenport, Iowa, no such remonstrances had any effect. People who sat within six or eight feet of the speaker soon discovered that they were not hearers. From all parts of the room came up the cry, "Louder! Louder!" Still the reading went on in a dreary monotone, without the slightest change in its pitch or force. Then members went up to the speaker and remonstrated; he waited til they had finished, and then went on in precisely the same tone as before. To observations from the chair he was as indifferent as to remarks less polite, but more forcible, from the body of the house. At last the fact dawned upon what cannot properly be called his audience, that Prof. Farquharson heard even less than they did. He was thoroughly deaf:
>
> Deaf to noun, and adverb, and particle,
> Deaf to even the definite article.
>
> There was nothing to be done under the circumstances but to let the reading proceed to the end. The real importance of the paper was such that Prof. Putnam read it before the Section again the next morning—this time so that it could be heard—and it justified the belief that it was of general interest. (*AAD*, *22*[2], 1877, p. 127)

Farquharson was deafened in the early 1850s while a young physician on the schooner *Taney* off the coast of Africa. W. D. Middleton, M.D. (1885), described the hearing loss as "in [Farquharson's] own estimation, so great an affliction, and which caused him to shrink from embracing so many opportunities of widening his sphere of usefulness, from a hyper-sensitive idea that communication with him was laborious and annoying" (p. 202). Farquharson resigned from the navy in 1855 and continued his practice in Nashville until the Civil War began. He was a fierce Unionist and a close friend of Andrew Johnson, who appointed him surgeon of the Fourth Tennessee Infantry. He served as head administrator for a hospital in Nashville, until 1868, when he traveled to Arkansas and, finally, to Davenport, Iowa, where he remained active in the medical community until his death in 1884.

THE DEAF PHYSICIAN AND THE BARRIER OF ATTITUDES

Middleton's 1885 biography of Farquharson reflects his belief that deafness greatly limited the physician's accomplishments. Given his many achievements, however, it is difficult to imagine how Farquharson could have accomplished more than he did. Like many highly competent people, Farquharson may have turned down opportunities, but he remained quite active in both scientific and public affairs. He served as president of the Iowa Academy of Sciences, presented numerous papers and published scientific reports, and, in 1880, he was elected a fellow of the American Association for the Advancement of Science, a reflection of his prestige.

This concern that deafness may hinder a person's success in a demanding profession such as medicine is commonly experienced by deaf medical students and physicians. It is documented in historical reports about deaf physicians, dentists, and nurses which illustrate this ongoing struggle to convince others that a deaf person can have a successful career in the health sciences. The prevailing climate in the nineteenth century, in particular, was generally not supportive of the notion of deaf persons entering medical professions. To some, deafness might have implied the inability to use good judgment, as in the case of the experience of the unnamed deaf professor at Montpellier mentioned briefly in the medical journal *The Lancet* in May, 1850. "One of the candidates for the chair of pathology and general therapeutics, at the faculty of Montpellier," the reporter explained, "lately challenged one of the members of the jury of professors, stating that he was incompetent to judge of the merits of the candidates, as he (the professor) was quite deaf. Though this infirmity is well known really to exist, the objection of the candidate was overruled" (p. 522).

In the editorial column of *The Lancet* in 1884, a disillusioned deafened physician posed to his colleagues the question whether his inability to hear would present a barrier serious enough to warrant a change of profession. After explaining his practice and the onset of his deafness, "Constant Reader" asked, "If you, Mr. Editor, or any of your readers, could tell me of a case where a man with very considerable deafness has yet continued to hold the confidence of his patients, it would be a comfort to me and an encouragement to hold on and persevere. If, on the other hand, it is to be my ruin professionally, then, devoted as I am to my calling, and little able, and still less inclined, at my time of life to launch out into another business or profession, I must look the evil in the face, and, for the sake of my children, commence the world, as it were, anew" (August 16, 1884, p. 307). This query by "Constant Reader" again illustrates to us the invisibility of deafness in this profession. Through the subsequent responses to his letter, several other nineteenth-century deaf physicians then practicing medicine have been identified. A month after this letter appeared, C. Corbett Blades attempted to provide some comfort by writing about his deaf father, who had practiced medicine for more than fifty years. He quoted his father's sister-in-law: "Your father never seemed to be at any loss with his patients" (August 17, 1884, p. 351). The following month another letter was received from E. W. Cushing, M.D., in Boston and published in this journal:

> Sir,—Allow me to inform "Constant Reader," in answer to his letter in your issue of Aug. 16th, that there is in this city a physician who carries a large practice although so seriously deaf as to be unable to follow ordinary conversation. The name of this gentleman can be furnished, if necessary. I have also heard of other cases where deafness seemed no bar to *retaining* an established practice. It would, however, doubtless prevent a young man from establishing *ab initio* such a practice. (October 11, 1884, p. 668)

Shortly after this, another deaf physician argued in favor of "Constant Reader's" continuing his medical practice: "I am one who can sympathise with your correspondent in his affliction . . . For five years I practised in the country in this colony, during which time I was very deaf, and had to use an ear trumpet. To this my patients took very kindly indeed" (November 29, 1884, p. 984).

In addressing the concern expressed by "Constant Reader," Dr. Cushing hinted at an attitude which prevailed—and still does—that congenital or early-onset deafness would "doubtless prevent a young man from establishing *ab initio* such a [medical] practice." Prevocationally deaf physicians in the nineteenth century were so few in number that the medical community might accept Cushing's reasoning in the absence of disputing evidence. (Proportionately speaking, too, the number of adventitiously deaf persons was higher in the absence of antibiotics and preventive vaccines.) Yet there is evidence of some acceptance by the medical profession of deaf physicians in principle, at least those with partial deafness, and this merits a closer look. In the March 1922 issue of *The Lancet* an advertisement for stethoscopes "for the use of deaf doctors" (p. 462) can be found, and in subsequent issues there are special attachments advertised for use with the telephone. And, in 1938, a stethoscope with a hearing aid was described which conveyed sounds to the ear by a microphone and a binaural attachment. The apparatus was more effective in amplifying heart sounds than breath sounds. The announcement claimed that this attachment "should be a great help to doctors who are deaf enough to need an electric hearing-aid" (p. 1211).

Up until this time no prevocationally deaf persons had entered a medical school. Ide L. Kinney began studying at Gallaudet College in 1887 and left after several years to take up the study of medicine. He became a licensed doctor of osteopathy, then moved into chiropody, where he became "signally successful" with "a list of over 3,000 satisfied patients, which is sufficient proof of his success [as] a foot specialist."[3] Edward C. Campbell also went on to a successful practice as a chiropodist. He was trained at the National College of Electro-Therapeutics in Indiana and developed the "Hygienic Sulfur Steam Bath Cabinet" for use in his large treatment center in Birmingham, Alabama. Neither Kinney nor Campbell earned a medical degree, however.

THE FOSSIL HUNTERS AND THE DARWINIAN REVOLUTION

Darwin's theory of "descent with modification" greatly increased the importance of taxonomic classification in identifying the diversities in nature, and the collection and classification of fossil specimens was invaluable for establishing a body of scientific knowledge. Darwin's books stimulated debates among geologists and paleontologists. Religious, philosophical, and scientific arguments both inspired and angered those who sought to unlock the secrets of nature. Among those who partook in the scientific quest and the religious discussions were several deaf scientists in Europe and America.

One of the first deaf European geologists on record was Ernest Grislet de Geer, the son of the mayor of Paris in the reign of Louis Philippe. He married a niece of a Swedish ambassador to the Court of St. James's and made his home in Geneva. In 1851 the "eminent geologist and antiquarian" made a tour of Scotland "to ferret out the old red sandstone strata and the other remarkable geological formations of the district [of Thurso]," during which he captivated deaf audiences with his graphic signs as he lectured (Roe, 1917, p. 357).[4] Sweden's distinguished geologist Alfred G. Nathorst (1850–1921), was deaf for much of his life. He was educated at the Universities of Uppsala and Lund and began his career with paleobotanical investigations of Mesozoic deposits. H. N. Andrews (1980) has written that Nathorst's deafness "may partially explain his voluminous correspondence" (p. 268). In 1870, he traveled to Spitsbergen to study phosphor beds as well as vegetation. In the following year he investigated glacial plant remains in Danish peat bogs and in 1872 he discovered the Arctic birch in peat at Mecklenburg, Germany. Nathorst received his doctorate from the University of Lund in 1873 and joined the Geological Survey of Sweden, remaining with it until 1884. He gained an international reputation for his paleobotanical investigations of the Tertiary flora. One of his most significant contributions was the improvement in methods for extracting information from fossil impressions. He described many important genera in his publications and authored several books, including *History of the Earth* (1888–1894) and *The Geology of Sweden* (1894), for which his followers were indebted to him.[5]

In England, Sir John Lubbock (later Lord Avebury) of the Royal Society employed a deaf microscopist and scientific illustrator, A. T. Hollick, to provide illustrations for his scientific publications. In 1872, Lubbock, a close friend of Darwin, published a contribution to natural history in which he discussed animals which generally had been neglected. Many were first grouped by Latreille under the name Thysanura. Lubbock, however, believed there were two distinct orders and proposed a second classification, the Collembola. The *Monograph of the Collembola and Thysanura* contained seventy-eight plates, thirty-one of them in color, beautifully illustrated by Hollick. The deaf illustrator's work was highlighted in the Deaf community in a 1902 issue of the *British Deaf Monthly*. He was a graduate of the school for deaf students in Brighton, England. Hollick's illustrations appeared in many publications, including those of Dr. Carpenter (author of *The Microscope*), Richard Owen, and Lubbock. He illustrated plates for the Royal Geological and Linnaean societies, and the records of the deep-sea expedition of HMS *Challenger* were completed by him in the form of lithographs. He also contributed many entomological illustrations for the Natural History Museum. In the preface to Lubbock's book it is explained that "the representations of the species, and the general execution of the plates, are the work of Mr. Hollick, a gentleman who is unfortunately deaf and dumb [sic], but in whom these terrible disadvantages have been overcome by natural genius."

> I believe this is the first work which has ever been illustrated by a deaf and dumb [sic] artist. It will be seen that Mr. Hollick has spared himself no labour or pains. I feel much indebted to him for the conscientiousness with which he has reproduced the minute details of the originals, as well as for the beauty and accuracy of his work. (J. Lubbock, 1872, p. vii)

In the United States, during the earlier decades of the nineteenth century, there were few libraries or even books to guide fossil hunters, and universities did not offer geology as a science. There were no railroads to carry scientists to the unexplored regions, nor to help them bring back their specimens of flora and fauna. New American discoveries were described in the established European journals as well as in the many publications which began to appear in the United States. Expeditions soon gave way to state-funded surveys in the 1830s, but political turmoil continued to slow the pace of these studies. During the Civil War, practically all scientific work was brought to a halt in the Confederate states and many records were destroyed. Only a few states were able to continue their geological surveys. Following the Civil War, American scientific work took a stronger foothold, particularly in astronomy, geology, meteorology, and botany. Agricultural and industrial use of natural resources grew exponentially, and large-scale manufacturing expanded throughout the nation. Railroads and telegraphic communication provided improved transportation and communication across large distances.

One deaf American adventurer of special note was George Catlin (1796–1872), an artist with wonderful talents in painting. Unfortunately, Catlin dabbled in geology as well. While traveling nineteenth-century America and producing portraits of Native Americans, he took notes on the striking features of the terrain, and published them. Merrill (1924) described Catlin's book, entitled *The Lifted and Subsided Rocks of America, with Their Influences on the Oceanic, Atmospheric, and Land Currents* (1870), as one of the two most "extraordinary" publications relating to American geology he had ever seen. "One can forgive any amount of ignorance relating to the subject of geology in a man of Catlin's profession," wrote Merrill, "but it is not so easy to forgive him for putting before an indiscriminating public opinions which are founded on wholly insufficient, and in many cases visionary, data" (p. 462). Catlin conceived that two large subterranean streams of water issued from both North and South America, one flowing south under the main axis of the Rocky Mountains, the other flowing north along the main axis of the Andes. This conjecture and his explanation of the origin of the Gulf of Mexico as being a result of the undermining of the earth's crust through the solvent action of heated water were accounted for, according to Merrill, "only on the grounds that he had received absolutely no training, had not learned how to observe, nor how to reason from that which he saw" (p. 462). The *American Journal of Science* also reviewed Catlin's book in 1870, dismissing his geological views as "consistent with the fact of his limited knowledge of the subject" (Merrill, 1924, p. 463).

Charles H. Sternberg (1850–1943) was one of hundreds of scientists who took to the mountains and prairies to pursue discoveries of fossilized plants and animals. In his autobiography, *The Life of a Fossil Hunter*, Sternberg (1931) described the problems and hardships encountered in fieldwork a century ago. Sternberg's adventures included expeditions to Kansas during which he endured attacks of malaria, "shaking ague," and long searches for drinking water while walking over miles of blistering chalk. He wrote of one humorous experience which occurred during an expedition to the Badlands, where his treacherous horse had no sensitivity to his hearing impairment:

> My right ear being totally deaf, I usually rode at the Professor's right, when the trail would admit of our traveling abreast. He was not always in a talkative mood, but when he began to speak of the wonderful animals of this earth, those of long ago and those of today, so absorbed did he become in his subject that he talked on as if to himself, looking straight ahead and rarely turning toward me, while I listened entranced. Not so that wicked black mustang of mine. Suddenly his front feet would leave the ground, and he would stand up at full length on his hind legs. Then feeling the gouging of the Spanish bit, he would drop and run ahead to the Professor's left side. When the Professor, happening to look up, found the place where I had been vacant, he would exclaim in surprise, "Why, I thought you were on my right, and here you are on my left!" The pony repeated this trick whenever I became so deeply interested in the Professor's talk as to loosen my hold on the reins. (pp. 69–70)

In his book, Sternberg described meeting the profoundly deaf paleobotanist Leo Lesquereux (1806–1889), whose work he had long admired. Lesquereux corresponded with Charles Darwin and published extensively on his own fossil discoveries. He was partially deafened as a child by a near-fatal fall from a cliff in his native Switzerland, and this was followed by a progressive loss of hearing. His mother became deaf in 1833 and his own deafness was then increasing. He suffered terribly from ulcers in his inner ear canal and was forced to give up teaching when he was no longer able to hear the recitals. To whom would he go for medical help? Dr. Jean-Marc Itard had by now built a strong reputation in France. Thirty years earlier, Itard was appointed by Abbé Sicard as physician in charge of attempting to educate the "Wild Boy of Aveyron." Itard had also demonstrated deaf pupils before the Academy of Medicine. His *Treatise on Diseases of the Ear* (1821) established him as a leading otologist. But among the deaf persons he "treated" he won few admirers. His callous attempts at cures

Leo Lesquereux, American Paleobotanist

for deafness included electrical stimulation, blistering, piercing of the eardrums, and fracturing of the skull behind the ear with a hammer, causing pain and death (Lane, 1984). Lesquereux was perhaps fortunate that he had not experienced a more serious fate when he submitted himself to Itard's treatment in Paris after corresponding with the physician. Lesquereux described the experience. The celebrated doctor, he explained, treated him shamefully:

> Demanding payment before making an examination of my case, fearing, he said, that I might leave Paris without paying his price, which was very high, he performed a first operation by liquid injections so strong as to produce an inflammation of the brain, and then refused to come to see me, saying that that was not his specialty, and that I must have another doctor. (Lesley, 1890, p. 198)

After immigrating with his family to America, Lesquereux established himself among the fossil hunters. He shunned meetings of the National Academy of Sciences as well as many other learned societies to which he belonged, preferring to work unobtrusively, although he was very capable of communicating with others in one-to-one or small-group situations.

Lesquereux's subsequent work led to a reputation as the nation's earliest authority on fossil plants. He was second only to William S. Sullivant in the field of bryology (the study of mosses). Sarton (1942) called him "the leading paleobotanist of the New World" (p. 97). He was the first elected member of the National Academy of Sciences and published many reports on his extensive analyses of fossil plants in the geological surveys of Arkansas, Kentucky, Indiana, Minnesota, Pennsylvania, and other states.

In 1863, Lesquereux informed his friend J. P. Lesley, "Darwin writes me that the few I have already published of the plants of the tertiary is very interesting" (Rodgers, 1944, p. 181). Lesquereux grappled for a long time with Darwin's thesis, reconciling the struggle for life with the "Providential law of development." Both Lesquereux and William Dawson, Canada's foremost expert on paleobotany, sought the opinions of Asa Gray, whose work was largely evoked by Darwin's. As with many American scientists who did not fully comprehend Darwin's *Origin of Species*, Lesquereux labored for a long time to gain a philosophical understanding of the theory in relation to his many investigations of geology and botany. "Religious man that he was," explained Rodgers, "Lesquereux had spent much time harmonizing the story of Genesis with the facts of scientific discovery" (p. 191). As Lesquereux himself said, "I know a *little*, other students of science know each a little, but the whole of what is known is but fragmentary and insignificant—merely a few pebbles picked up along the ocean shore" (p. 191).

By 1864, Lesquereux had begun to show more comfort with the religious implications of Darwin's work: "I have studied and studied again Darwin's . . . origin of species [sic] and the more I read it the better I am pleased with it. This system explains to me some of the mysteries of our Christian revelation which had been obscured to my mind till now. For to tell you the truth I read the Bible every day and constantly find it new light and new life" (Rodgers, 1944, p. 19).

And by 1873, as Rodgers wrote, "Lesquereux foresaw science pointing the way to a more truthful theology, and religion encouraging scientific research" (p. 19).

L. R. McCabe (1887) interviewed Lesquereux in his later years and asked him if his long and intimate associations with such great men as Schimper, Brongniart, Marquis Gaston de Saporta, Schenk, Williamson, and Nathorst had not left him with many anecdotes and fond memories. Lesquereux remarked:

> The science-student's life is absorbed with grave and serious truths; they are naturally serious men. My associations have been almost entirely of a scientific nature. My deafness cut me off from everything that lay outside of science. I have lived with Nature, the rocks, the trees, the flowers. They know [me], I know them. All outside are dead to me. (p. 839)

This comment by Lesquereux aroused my curiosity, for it was only through the research for this book that I learned of the intimate association a number of his colleagues had with deafness, yet Lesquereux made no mention of this. Alfred G. Nathorst was deaf. William Williamson also had a persistent "ear ailment" which required special medical attention. Lesquereux also worked with Sternberg and Fielding Bradford Meek (1817–1876), the former deaf in one ear and the latter progressively deafened like himself. Had these men ever put down their fossils and discussed their common experiences with deafness?

Meek's hearing began to fail when he was young and progressed until he was totally deaf. It is possible that a congenital syphilis infection was responsible for the hearing loss. In a letter written to J. S. Newberry, Meek described his desire to isolate himself from the demands inherent in communicating with others: "I also prefer to spend my evenings with the books and specimens. I can hear and understand what they say, and they require neither small talk nor formalities" (Merrill, 1924, p. 528).

According to Charles A. White (1896), a physician-turned-paleontologist with a longtime connection to the U. S. National Museum in Washington, D.C., Meek never developed skills in speechreading. White described his friend as an introvert who seldom complained about his deafness. He compared Lesquereux and Meek, the latter never having learned how to "converse orally by watching the motion of the lips of the speaker":

> [Meek also] seemed averse to the use or recognition of any conventional signs. Still, he was always ready and eager to converse with his friends, and he always kept at hand a pad of paper and a pencil for their use. Although he was thus cut off from the oral conversation of his friends he never referred to it as a hardship. Because of this affliction, however, and of his natural diffidence, he rarely, if ever, attended social gatherings. (pp. 341–342)

The touching biography of Meek published in *The American Geologist* in 1896 was derived from White's interview with his deaf friend. It begins with Meek looking up from a collection of fossils the two of them had been classifying together for the Smithsonian Institution. Meek told White about a dream he had the night before in which he had heard his sister singing a song while playing the piano. "She has been in her grave for thirty years," Meek said, "and my deafness is so complete that I have not heard a sound of any kind

Fielding Bradford Meek, American Geologist

for a long time" (White, 1896, p. 338). Meek was not one to be personal in his conversation, and White quickly grabbed the opportunity to encourage his friend to talk more about himself. This interview with Meek and written communication between Meek and James Hall are the primary sources of information I have found. Meek died two months after his dream.

It is remarkable that while suffering from continued health problems he was able to travel extensively and live under the most challenging conditions. In Merrill's (1924) book *One Hundred Years of American Geology* is found an appendix with a series of letters Meek wrote while traveling by steamboat up the Missouri River, in which he described his encounters with Native Americans and other fossil hunters on his way to the Badlands. In addition to his deafness, Meek had pulmonary tuberculosis during much of his life. Yet even with these challenges, few others have published a larger quantity of descriptions of invertebrate fossils. Much of his work remains valid today. He discovered new genera and subgenera, species of Acephala from Cretaceous formations, Gasteropoda from the Tertiary, and others from the Jurassic and Paleozoic periods. He studied species from nearly every phylum, and coauthored many papers with F. V. Hayden, a lifelong colleague. Meek's last publication alone, Volume IX of the *United States Geological Survey of the Territories* (1876), contained more than 600 pages and forty-five illustrations. He was honored with membership in the National Academy of Sciences and, upon his death, his friend Joseph Henry delivered the funeral oration in the Smithsonian Institution hall.

In addition to Leo Lesquereux, another deaf man who contributed to the work of Charles Darwin was Thomas Meehan (1826–1901), an American horticulturalist. The early literature of the Deaf community occasionally mentioned men and women who won various prizes and other honors for their work with flowers and plants. They were primarily gardeners and hobbyists, like Sarah F. Lewis of Yorkshire, England, and Arthur L. Harvick and Paul Lange in the United States. In *The Botanists of Philadelphia and Their Work* (1899), Harshberger described three deaf botanists whose work was probably not known in the Deaf community. Benjamin M. Everhart was a specialist in the study of fungi. Aubrey H. Smith, a friend of General Grant, presented an extensive herbarium to the Biological School of the University of Pennsylvania. Some of Thomas Meehan's work was of value to Charles Darwin and was cited in his publications. Both Everhart and Smith lost their hearing adventitiously, while Meehan was congenitally deaf. Born at Potter's Bar, near Middlesex, England, Meehan spent his childhood on the Isle of Wight where he learned gardening from his father. Deaf and in a family of deep poverty, he was taught by his mother until he was twelve, when he began to work as a gardener. Harshberger wrote that "being deaf from birth, [Meehan] never mingled with other boys, but spent his time in the fields as an amateur naturalist. Some of his early observations were remarkable, and attracted the attention of well-known men, who befriended him" (p. 249).

In *American Botany* (1944), Rodgers described Meehan's explorations of the North Carolina, California, and Colorado mountains. Harshberger (1899) reported on his work in Alaska. Meehan published over 200 scientific papers and served on the Harvard University Board of Visitors. He advocated the

Thomas Meehan, Dean of American Horticulture

establishment of small parks and was largely responsible for the formation of the City Parks Association in Philadelphia.

Harshberger (1899) has summarized Darwin's use of Meehan's work:

> Professor Meehan, as a scientific man, has corresponded with most of the scientists of prominence in both Europe and America. A close correspondence was maintained with Charles Darwin, who relied on Mr. Meehan's observations for many of the facts incorporated in his books. This correspondence continued, until a slight misunderstanding between them finally put a stop to their letter-writing and pleasant intercourse. Mr. Darwin gives credit to Meehan's acute observations in many places in his epoch-making works. (pp. 253–254)

Darwin's correspondence provides details about the nature of some of Meehan's contributions as well as the source of their disagreement. As early as June 10, 1862, for example, Darwin wrote to Asa Gray seeking Gray's opinion on a paper Meehan published in the *Proceedings of the Philadelphia Academy of Natural Science* in which Meehan compared species of European and American trees. Darwin was impressed with this work. Ten years later, however, Darwin received a letter from Hubert Airy (September 20, 1872) in which Airy disputed Meehan's observations on the hardiness of exposed buds. On October 9, 1874, Darwin wrote to Meehan and encouraged him to continue his investigations of the colors of dioecious flowers: "Some thirty years ago I began to investigate the little purple flowers in the centre of the umbels of the carrot. I suppose my memory is wrong, but it tells me that these flowers are female, and I think that I once got a seed from one of them; but my memory may be quite wrong. I hope that you will continue your interesting researches" (Darwin & Seward, 1903).

The following year, 1875, Meehan published a report in which he denied the importance of fertilization of plants by insects. Hermann Müller wrote to Darwin encouraging him to work on this. Over the next few years, Darwin and Meehan had a falling-out. When Meehan sent one of his more recent papers on this topic in July 1877, he suggested that Darwin and he now agreed on the extent of self-fertilization in nature. Darwin, however, still did not concur with Meehan's work and responded within a few days with a letter in which he wrote that he feared "we must agree to differ!"

In 1883, at the age of fifty-seven, Meehan presented a paper entitled "Variations in Nature," in which he supported Darwin's theory. Earlier, the American Association for the Advancement of Science had invited him to occupy the chair of one of its sections, but Meehan declined, "fearing that his imperfect hearing would interfere with the proper performance of the duties of the chair" (Meehan, 1883, p. 437). In his paper on variations, he wrote:

> To my mind the doctrine of the origin of species as presented by Mr. Darwin cannot be controverted; and the theory of natural selection which he elaborated to sustain this doctrine requires only to be pruned of some abnormal growths to make it impregnable. This has been my task during the years referred to. Whenever I have seen weaknesses, I have adduced facts to oppose them. My task to-day [sic] is, as I hope, to show that the

doctrine of evolution, and especially the theory of natural selection, is all the stronger for the work I have done. (Meehan, 1883, p. 437)

Meehan discussed how his own research had generally supported Darwin's theory, but he questioned Darwin's view of variability in nature as a slow process and his contention that the variations are always in the "line of profit"—that is, survival of the fittest. He concluded his presentation with a firm statement in support of science being "allowed to work out its own salvation, if there be aught in it worth saving" (p. 447). "There is no conflict between the doctrine of evolution with its theory of natural selection and the essential doctrines of Christianity," he told the audience. "Rather, as I believe, will the day arrive when those who advocate these views will be regarded as Christianity's best friends" (p. 448).

For his lifelong work on the study of plants, Thomas Meehan has been honored with the title "Dean of American Horticulture."

THE ESTABLISHMENT OF GALLAUDET COLLEGE

Gallaudet College came into being because of both attitudinal barriers and a consequent need to overcome them. In 1857, Amos Kendall, the business manager for Samuel F. B. Morse, met with Edward Miner Gallaudet, the son of Thomas Hopkins Gallaudet, and encouraged him to accept responsibility as the superintendent of a school for deaf and blind children which Kendall had established the previous year in the District of Columbia.[6] The Columbia Institution for the Deaf, Dumb and Blind, incorporated by Congress in 1857, was authorized by President Lincoln in 1864 to grant college degrees in the liberal arts and sciences.[7] The first freshman class was admitted to the collegiate program, the "National Deaf-Mute College," in September of that year. At the inauguration of the college on June 28, 1864, the eminent deaf educator Laurent Clerc praised science as "a most useful thing for us all," but in this particular address, Clerc lacked total conviction that deafness should pose no barrier to the aspiring deaf person: "The degree of Master of Arts can be conferred on the deaf . . . when they merit it; but, on account of their misfortune, they cannot become masters of music, and perhaps can never be entitled to receive the degree of doctor in divinity, in physic, or in law" (Fischer & de Lorenzo, 1983, p. 227).

The distinguished authorities with whom Edward Miner Gallaudet was acquainted included Professor Joseph Henry, the first secretary of the Smithsonian Institution, and the Arctic explorer General Adolphus Greely.[8] Gallaudet's friendships were to his political advantage. One of Gallaudet's responsibilities during his early years at the Columbia Institution, for example, was the planning of suitable courses of study for the college. But Kendall, dominating the board, offered his own plan which, in Gallaudet's words, "would have made the college a laughingstock of all educated men" (Boatner, 1959, p. 38). The conflict between Gallaudet and Kendall led Gallaudet to seek support from his close friend, Henry, of whom Edward wrote, "no higher authority among educated men could be found" (Boatner, 1959, p. 39). Henry, one of the greatest experimental physicists in American history, was appointed by President Lincoln as one of the original fifty leading scientists to form the National Academy of Sciences in 1863, and he served as its president until his death. Few educators of deaf students and even fewer physicists know of Joseph Henry's friendship with deaf scientists and his interest in the education of deaf students. It is uncertain how he developed this interest, but it probably began with his initial associations with deaf scientists. His involvement with the Columbia Institution deepened as his friendship with Edward Miner Gallaudet grew. Henry was only too happy to assist Edward in his dispute with Kendall over the school's curriculum. In his supportive letter to Gallaudet on November 16, 1864, he praised the school, explaining that "the Smithsonian Institution affords unusual facilities for the study of science," and that "as director of this establishment I am authorized to say in behalf of the Board of Regents that these facilities will always be free to the pupils of the Columbia Institution" (Fischer & de Lorenzo, 1983, p. 41).

Kendall must have listened to Henry's advice, as Gallaudet noted: "To my great surprise, when the Board next met, Mr. Kendall did not call up his regulations, and he never again alluded to them to me in any way" (Fischer & de Lorenzo, 1983, p. 43).

Henry also spoke at the college on Presentation Day in 1876:

> I am now happy to say that the experiment has been successful. . . . The deaf . . . from the want of the sense of hearing are capable, perhaps, of more undisturbed attention and of sustained effort for the accomplishment of high mental achievement. They are specially well adapted to various scientific investigations, and may be successful laborers in the line of astronomy, heat, light, electricity, magnetism; in the great domain of chemistry and natural history; in short, in every branch of human thought. . . . Why should they not, therefore, be educated to the full extent of their mental and physical capacity? . . . That [the Columbia Institution] continue to prosper and bring forth fruit commensurate with the object of its foundation is my most earnest prayer. (Fischer & de Lorenzo, 1983, p. 124)

Henry's optimism about careers in science for deaf persons is understandable. At the very moment he spoke on Presentation Day in 1876, there lived in a wing of the Smithsonian Castle the deaf paleontologist Fielding Bradford Meek, who had distinguished himself as a foremost expert in his field. Henry's interest in geology had kept him in contact with Leo Lesquereux. One of the students who graduated at the first commencement of Gallaudet College, James H. Logan, was a talented microscopist whom Henry had helped to purchase instrumentation for his research. In 1857, Henry also had served with Frederick Augustus Porter Barnard on an American Association for the Advancement of Science "Committee of Twenty" to inquire into the Coast Survey of the United States. Among scientists, Joseph Henry was no stranger to deafness.[9]

By the 1870s many of the institutions for deaf pupils in the country had established "High Classes," but they were still grappling with the issue of providing quality science education. F. D. Clarke argued eloquently in an article "Science for the Deaf and Dumb":

> We may not be able to make of our pupil a Kepler, a Newton, or a Tyndall; though there is no reason, except want of proper education, why deaf-mutes [sic] might not think out the laws of the planets, analyze light, or determine the radiation of heat through aqueous vapor, and the relation of that to the climate of the earth. But, if they cannot look through the telescopes to discover new worlds in the heavens, or through the spectroscopes to find new constituents in the flame of a candle, they can direct our mines, engineer our railways, preside over our chemical works, assay our metals, reduce our ores, plan our bridges, take our photographs, make our microscopes, be our geologists, our botanists, our surveyors, our architects, and our engravers. (*AAD*, *16*[2], 1871, p. 101)

How well was Gallaudet College preparing deaf students in science during these early years? One of the first evaluative comments addressing this issue came from Joseph Henry in 1876 when, after reviewing the education of the deaf students, he stated emphatically that "the scientific examination papers of last year were submitted to me for report as to their character; while they involved the solutions of questions in mathematics, physics, chemistry, geology, etc., requiring accurate knowledge and profound thought, *the answers were such as to do honor to the undergraduates of any college in this country*" (Fischer & de Lorenzo, 1983, p. 122). But not all of the graduates would agree that the facilities were adequate. As William L. Hill, who graduated in 1872, testified in his "Glimpses of the Past: Reminiscences of the Sixties" (Hill, 1895), science-minded students in the early years at the college were rich with excellent teachers, but poor in laboratory materials and facilities:

> We had some very elucidating experiments in Chemistry. Prof. Spencer carried around the laboratory in his coat-tail pockets, and sometimes showed us a vial of stuff which he gravely assured us would, if combined with some other kind of liquid (which we were invited to go and gaze upon at a drug store or the Smithsonian) produce some very astonishing results. . . . We studied Mineralogy and Botany and Astronomy in the same captivating and elucidating way. The professors did the best that could be expected under the circumstances, and we doubtless profited by their teachings materially, but the poverty of our resources in the matter of apparatus, cabinets, etc., told constantly against us, and the most of the learned and eloquent disquisitions of our faithful preceptors fell, literally and figuratively, upon unheeding ears. When last spring I visited the College and beheld its superb chemical laboratory, its cabinets of minerals, its astronomical charts and globes, its splendid maps, its fine library and its many other resources for thorough practical instruction, a pang of regret smote me, and I felt deep down in my chest that it would have been money in my pockets if I had been born deaf, or achieved deafness, or had deafness thrust upon me twenty years later. (p. 62)

A look at the course of study for deaf students in 1893 indicates that science and mathematics were an integral part of the required program. Freshman were expected to take a full year of mathematics, including algebra and geometry. Sophomores studied plane and spherical trigonometry, analytical geometry, zoology, botany, and chemistry. In the junior year, calculus was optional, and

First Electricity Laboratory Building in 1904 at Gallaudet University

physics (including astronomy), chemistry, and physiology were required. The final year included courses in natural science and psychology (Gallaudet, 1893). By the turn of the century, however, there were still very few deaf scientists known to the Deaf community. When James E. Gallaher highlighted the successes of deaf people in his book *Representative Deaf Persons of the United States of America* (1898), he developed 147 biographies in his first edition. Only three were scientists. In preparing for the second edition, also in 1898, he received 850 names of "representative deaf men and women" from superintendents and other contacts in schools for deaf children in the United States. Only 201 responded, none of them scientists, and Gallaher expressed his disappointment with the results of his survey: "The American deaf were given a chance to be represented in a book in great numbers, so that the world could see what they have accomplished and are doing, but they failed to appreciate what was offered them" (p. 3). Given the typical response rate for such surveys, modern-day researchers would likely say that Gallaher's disappointment was unwarranted.

Gallaudet College's Impact on Science as a Profession for Deaf People

Although primarily a liberal arts college, during its early years Gallaudet College provided an important stepping-stone for deaf students inclined toward occupations in science. Prior to its establishment, there was little accessibility for deaf persons to higher education at all. Now, with a college of their own, and with instructors capable of communicating in sign language, additional deaf American scientists were being trained. One of the three men who graduated at the first commencement of Gallaudet College in 1869 was James H. Logan (1843–1917), deafened at the age of four and a half from scarlet fever. In the first of a series of letters now kept at the Smithsonian Institution, where Joseph Henry was then secretary, Logan wrote to the distinguished physicist and requested assistance in acquiring a Smith and Beck monocular microscope from Europe "to aid me in studies to which I am specially devoting myself." Henry's kind efforts to get the young deaf Gallaudet student started paid off. In time, Logan's scientific interests led him to a patent for a microscope, published in the July 1869 issue of *Scientific American*. He also played a primary role in the founding of the Iron City Microscopical Society in Pennsylvania. Later, he supervised a staff at the Department of Agriculture in Pittsburgh and held a position as "demonstrator" in biology and zoology classes at the Western Pennsylvania Medical College.

I planned to study back issues of *The Western Pennsylvanian* for additional information about Logan's life, since he held the position as the first superintendent at my alma mater, the Western Pennsylvania School for the Deaf (WPSD) near Pittsburgh. After describing my research on deaf scientists to James Salem, William Craig, and Harold Mowl, administrators at the school, they took me to the board of directors conference room to show me a beautiful brass telescope which had been recently moved there. It was in 1989 and the faculty and

James H. Logan, American Microscopist

students were curious about who had built the instrument and how it had arrived at the school. Since the early decades of this century, the handmade telescope had stood in a back room, occasionally inducing curious but unknowledgeable comments. Determined to reconstruct its origin, I searched through *The Western Pennsylvanian* and found that the telescope had been designed and built by Frank Ross Gray (1856–1924), a deaf lens maker at John A. Brashear's optical company in Pittsburgh. Gray graduated from Gallaudet College in 1876. To students interested in astronomy at the Western Pennsylvania School for the Deaf, he was a man who gave freely of his time. He offered them his knowledge of the heavens, never expecting anything in return, and he instructed them in the use of the instrument. In Gray's obituary, Vincent G. Dunn (1924) wrote that "He knew the starry heavens so intimately that it was a pleasure to him to render such service" (p. 356).

In 1885, Logan helped to bring Gray to the Pittsburgh area, where he accepted the position with Brashear. With the financial assistance of philanthropist William Thaw, Brashear's firm had begun to prosper. The quality instruments the company produced included refractors, reflectors, and Bruce doublet wide-field cameras for observatories in Victoria, Johannesburg, Heidelberg, and in the United States at Yerkes Observatory and the nearby Allegheny Astronomical Observatory in Pittsburgh. As the industrial scene in the United States changed, the "Little Paper Family" began to report a number of deaf people entering lens making as an occupation. No deaf lens maker, however, directly contributed to the field of astronomy as much as Frank Ross Gray at the Brashear Company. Gray made over 50,000 lenses and prisms in his career. He was commissioned by the Canadian government to work on a six-foot astronomical mirror in 1915. Having worked for Brashear on Conservatory Hill for over thirty-five years, he and "Uncle John" were very close friends, and Brashear paid Gray a great honor in naming him as one of his pallbearers in "the grandest funeral Pittsburgh had yet given any man" (*B&B*, *28*[8], 1920, p. 295). Gray would be happy to know that his telescope is still at WPSD, where he had inspired so many friends to look upward into the heavens and enjoy the wondrous canopy of stars. As Brashear once said, "Somewhere beneath the stars is work which you alone were meant to do. Never rest until you have found it" (Brashear, 1988, p. 28).

A third early Gallaudet alumnus to become a scientist was George T. Dougherty (1860–1938), an excellent example of the benefits realized from the establishment of schools for deaf students in the United States. Dougherty was head chemist and metallurgist at the American Steel Foundries Company. He published widely on his research on the determination of nickel in iron and steel, salt in petroleum, and the process for converting oils and fats into soaps. His accomplishments were notable considering that his primary communication mode with hearing people was writing. In 1912 he remarked, "Lip-reading in volves too much guess-work. Writing is far safer and more dependable in social and business intercourse on the part of the deaf with hearing people, or *vice versa*" (Braddock, 1975, p. 202).

Born deaf, Dougherty remained intensely involved in the Deaf community throughout his life. While a student at Gallaudet College, he helped to found the

George T. Dougherty, American Metallurgist

National Association of the Deaf (NAD), organized at Columbus, Ohio, in 1880 and he served as its first secretary. He also held the office of NAD vice president, and, many years later NAD elected him a life member. In 1893, Dougherty served as chairman of the World Congress of the Deaf in Chicago.

Dougherty's involvement with the preservation of American Sign Language (ASL) well illustrates his dedication to the Deaf community. As with other languages, ASL has a rich history. Since the turn of the century, various efforts have been made to preserve early examples of ASL, especially by the NAD. In one of the first motion picture films produced for this purpose, the deaf chemist George T. Dougherty provided an eight-minute presentation, "The Discovery of Chloroform."

During summer vacations from Gallaudet, Gerald M. McCarthy (1858–1915) hiked through the mountains and marshes of North Carolina and South Carolina, having one adventure after another as he collected ferns and other specimens. Some he gave to the Smithsonian Institution, whose officials began to recognize his worth, and between 1887 and 1889 McCarthy helped them to name specimens. Other specimens were purchased by botanical societies in Europe. Upon graduation from Gallaudet in 1887, McCarthy was immediately offered a position at the recently established Agricultural Experiment Station in Raleigh, North Carolina. Over the next year he published several scientific articles and, with Thomas F. Wood, coauthored a book entitled *Wilmington Flora: A List of Plants Growing about Wilmington, North Carolina, with Date of Flowering, and a Map of Hanover County*. In *American Botany* (1944), A. D. Rodgers recognized McCarthy's contributions: "Lieut. V. Havard's 'Report on the Flora of Western and Southern Texas' and Gerald McCarthy's 'A Botanical Tramp in North Carolina' also added to exploration knowledge of the time; as did [Asa] Gray's own account of his and John Ball's excursion to Roan Mountain in 1884" (pp. 232–233). As secretary of the North Carolina State Horticulture Society he published numerous reports on his research on insects and diseases. His description of the history of silk production led the North Carolina Agricultural Experiment Station to request that he conduct a more extensive study. McCarthy served as both state botanist and state entomologist for North Carolina. "For several years before and after the turn of the century," wrote Gilbert Braddock, "the deaf made prideful mention of their representative in the learned fields of botany and entomology—the only deaf man in the world properly qualified by study and training to identify any leaf or flower or any bug or beetle appearing in any given locality" (1975, p. 179). Yet, with all of these accomplishments, he still faced constant battles for fair wages and was forced to file a lengthy complaint against the North Carolina Board of Agricultural Trustees and the North Carolina Board of Health for salary due him.

Considering the fact that several faculty members at Gallaudet College, including its second president, Percival Hall, and a vice president, Charles Russell Ely, had strong interests in entomology, it is surprising that more Gallaudet students did not pursue this field of study. Perhaps this was because Gallaudet's curriculum was much stronger in chemistry. One student, Charles F. Neillie, never completed his education at Gallaudet, leaving in 1891, but did become the city entomologist in the Forestry Division of the Department of

Gerald M. McCarthy, State Botanist and Entomologist of North Carolina

Public Service in Cleveland, Ohio. Neillie pursued entomological studies throughout his career, developing an aircraft insect spray technique and investigating insect blights affecting trees. Another graduate, Cadwalader Washburn, deaf since the age of five from both scarlet fever and spinal meningitis, became an eminent artist, but also contributed to the field of entomology. Washburn began his studies in entomology at Gallaudet College, where his interests "awakened in him a desire to be able to execute pictorial illustrations of his essays on insects" (*SW*, *10*[5], 1898, P. 67). His investigations and essays in entomology, and the collections of insects made during his holiday excursions into the Washington countryside, attracted the attention and commendation of the Gallaudet faculty. One of Washburn's most important papers on this subject was his senior dissertation, "The Working Mind of a Spider," which he read in sign language at his commencement. Other papers for which he was honored include "Some Experiments with a Chrysalis" and "Language of Bees."

Without doubt, the impact of Gallaudet College in the United States was incalculable. Their Gallaudet experience influenced not only the success of these early graduates in science, but also helped shape them into advocates for continued improvement in the educational and employment opportunities for other deaf people.

BREAKTHROUGHS IN EUROPEAN HIGHER EDUCATION

Meanwhile, deaf people in Europe had no higher education opportunities comparable to Gallaudet College. When the first profoundly deaf youth educated in institutions for deaf children succeeded in the university systems of Europe, the respective Deaf communities took special pride. The first person to do so in England was Abraham Farrar (1861–1944), who received many accolades in 1876 when he passed the Cambridge University Local Examinations at the age of sixteen. Farrar had been carefully tutored for thirteen years by Reverend Thomas Arnold of Northampton. In reporting on Farrar's obtaining a university B.A. degree, the British Deaf community emphasized the fact that he was "absolutely deaf." He received personal congratulations from such eminent Victorians as the Prince of Wales (Edward VII) and Earl Granville, the chancellor of the University. Farrar's interest in science continued for many years. "Ever since boyhood," he wrote in an autobiographical sketch in 1939, "I have been interested in science, and for a time was interested in botany; but afterwards took up the study of geology in which I had expert guidance" (pp. 27–28). One of Farrar's papers was read for him at a local meeting of the Leeds Geological Association in northern England and he was elected a fellow of the Geological Society of London on January 7, 1898. In the years to follow, he attended excursions with the Geological Association, "the most memorable of which was one to the wonderful extinct volcanoes of Auvergne in Central France" (Farrar, 1939, p. 28). Farrar later turned to education and published many articles on the subject of deafness.

In France, the first deaf person to earn a bachelor's degree became a chemist. Ernest Dusuzeau had studied in the Imperial Institution in Paris, and the French

Deaf community praised him for overcoming the seemingly insurmountable odds when he graduated from the Sorbonne in 1885. His educational attainments in this period of history brought honor to his former teachers. Dusuzeau is known more for his political and educational advocacy than for his work in chemistry. He was profoundly deaf and a strong proponent of the rights of deaf people both in France and in America. At this time, deaf people were fervently reacting to the banishment by hearing people of sign language in the schools. Unable to teach articulation, he was one of the deaf instructors forced into retirement from the institution at St. Jacques in Paris by the "Edict of Milan" at the 1880 International Congress on the Deaf (see Chapter 3, "The War of Methods"). He was humiliated by the loss of his position and he rallied the leaders of Deaf communities of various nations to defend their right as human beings to communicate as they desired. "There have been many congresses aimed at improving our lot," Dusuzeau argued, "but none of them has been satisfactory." He believed speech to be a great gift, but that in encouraging deaf people to learn speech, the use of sign language should not be banned.

DEAF WOMEN: THE STRUGGLE FOR ACCESS

Only a few years after the first class graduated from the National Deaf-Mute College in Washington, D.C., Laura C. Sheridan (1875) presented a passionate plea for higher education for deaf women: "The world has lost immensely by being so long in awaking to the importance of equal education for woman [sic]," she wrote (p. 249). Sheridan was a deaf teacher at the Indiana School for the Deaf. She enrolled in a correspondence program of the Chatauqua Literary and Scientific Circle and received a diploma. In her 1875 effort to bring the plight of deaf women to the attention of authorities, she wrote: "So there has been much agitation of the question of the higher education of woman within the last few years, the result of which is that the doors of colleges and universities are opening to her everywhere. But what have we heard of the question in the silent world? Nothing . . ." (p. 248).

Indeed, nothing happened for another six years until January 11, 1881, when the issue of admitting deaf women to the college was discussed at a faculty meeting. "The Faculty," observed Edward Miner Gallaudet, "showed no disposition to change the policy of the College which declines to admit ladies" (Boatner, 1959, p. 114). In the course of working hard against one form of discrimination, Gallaudet and his colleagues were actively defending another equally obvious form. The accomplished poet and deaf feminist Angeline Fuller Fischer even called for a separate college for deaf women, but Gallaudet did not change his opinion. Three years later he again expressed this opposition while visiting a school for deaf pupils in Council Bluffs, Iowa.

Continued pleas urging the college authorities to admit women were presented at the national conventions of instructors of deaf pupils, but the faculty did not drop its resistance until 1887, when it opened the college doors to women—as an "experiment." In this year, the first deaf women students were admitted "on the same terms as the men."

The college admitted six women in the fall of 1888. The number of women was restricted in part due to housing, a concern which had been one of the reasons Edward Miner Gallaudet had resisted coeducation from the start. This problem was solved by Gallaudet, who was considering retirement, when he offered to vacate his home during the "experiment." In 1916, Congress passed an act to provide funds for a new dormitory for women students, which was named in honor of Edward's mother, Sophia Fowler, who was deaf.

Deaf women were expected to take the same program of instruction as men, including the range of courses in science, mathematics, and astronomy. An examination of the college records indicates a promising time when these early women students gained their long-sought equality of opportunity. The first deaf woman to graduate from the National Deaf-Mute College was Alto M. Lowman of Maryland, who received a bachelor of philosophy degree in 1892. In 1893, Agatha Mary Agnes Tiegal, the second woman graduate of the college and a former student of the Western Pennsylvania School for the Deaf, reflected the spirit of the day in her Presentation Day oration, "The Intellect of Woman." She was the first deaf woman to receive a bachelor of arts degree. Emma R. Kerscher, also from Pennsylvania, presented an oration four years later titled "The Achievements of Woman," and Emma Hall's oration in 1905 was "The Martyrs of Science." Gallaudet College's first deaf woman instructor, May Martin, accepted a position teaching English after she graduated in 1895. It was not until 1961, nearly a century after the college was authorized to grant college degrees, that a deaf woman science professor was hired at Gallaudet College. She was Edith Rikuris, a Gallaudet graduate in 1961 who earned a master's degree from Catholic University.

That deaf women were not entering science was hardly surprising. Hearing women were also struggling for acceptance and recognition in scientific professions. There were no visible models for young deaf women to emulate and few deaf women science and mathematics teachers in the schools. And, importantly, Gallaudet College itself had not yet begun to accept deaf persons in the "Normal Department," which trained such teachers. In fact, no deaf student was admitted as a candidate for the graduate degree in education at Gallaudet College until 1963.

At the turn of the century, the issue of employment of deaf women gained in importance in the Deaf community. National conventions and publications addressed the question of why deaf women who had earned degrees at Gallaudet College were not entering professions. In a paper on employment opportunities for deaf women presented at the Sixth Convention of the NAD in St. Paul, Minnesota, May Martin described the obstacles. "It seems, after all," she explained, "that there are no gates absolutely closed to a deaf woman, but I would not arouse nor encourage false hopes; there is much prejudice to overcome; ignorance to struggle with; many employers know too little about the deaf, and the capabilities of the deaf, to be willing to give them work, or to trust them with the kind of work they desire" (Martin, 1899, p. 81).

Ten years later, an article titled "A Gentle Appeal to American Deaf Women" was published by a writer named "Pansy" (later identified as Gertrude Maxwell Nelson):

> It does seem strange, yet it is a most noticeable fact that many deaf women graduated from Gallaudet College; there are very few, indeed, who have [cultivated] a taste for literary work or done anything for their own sex that could be recorded as noteworthy. We would very much like to know what these women are doing with the fine education they have acquired, and which is much above the average obtained at State schools—are they putting these rare attainments to any good purpose by hiding them under a bushel? (*SW*, *21*[9], 1909, p. 159)

J. A. Sullivan (*SW*, *36*[7], 1924) conducted a short follow-up study of Gallaudet College graduates for the years 1913 to 1922. During this period, seventy-one men and forty-seven women earned degrees from Gallaudet College. Of the women graduates, three had died and twenty-seven were married and not in the workforce. Seven were teachers (one in mathematics, none in science). The occupations of the remaining ten were not related to scientific work. A comparison of these data with the occupations of the male graduates is revealing. Nine of the seventy-one males during this period entered occupations associated with science and mathematics. Of these nine graduates, two were instructors at Gallaudet College in mathematics and chemistry, six were chemists (three of them having received master's degrees at McGill, George Washington, and the University of California), and the last was a bacteriologist working for the Canadian government.

Although more than a third of the bachelor's degrees awarded by Gallaudet College went to deaf women between 1922 and 1935, the improved educational opportunities did not lead to any observable change in the number entering science professions. The periodicals of the Deaf community, the most promising source of information, are devoid of any reports of deaf women scientists. As far as I can discern, the first deaf woman to attend Gallaudet College and subsequently engage in scientific work was Katharine M. Schwartz, who left the college in 1903. Schwartz completed a B.A. degree at the University of Minnesota in 1910 and published a report titled "With an Expedition for Scientific Research" (*B&B*, *27*[5], 1919, pp. 167–170). In this article she described her travels with a group of scientists investigating the vegetation of the western United States, particularly the arid regions. All of the members of this group were associated with the University of Minnesota. During these early years there were few deaf women attending colleges and universities with hearing students and no evidence that any congenitally deaf women entered scientific studies. Hypatia Boyd studied science at the University of Wisconsin in 1895, but was forced to withdraw when her father experienced financial difficulties. Boyd obtained a position as a writer for one of the leading daily papers in Milwaukee.

The first college for women in England was located three miles from Cambridge University. Girton College opened in 1869, and some of the professors at Cambridge allowed the women to attend their lectures. Male students graduated from Cambridge with honors by taking the Tripos examinations. Patricia Clark Kenschaft (1987) described how, in 1880, Charlotte Angas Scott (1858–1931), after more than fifty hours of the Tripos mathematics examinations, had done as well as the eighth man at Cambridge.

The news spread quickly through all of England that a woman had succeeded in a "man's subject." As Kenschaft explained:

> Because she was female, [Scott] could not be present at the award ceremony, nor could her name be officially mentioned. However, a contemporary report says, "The man read out the names and when he came to 'eighth,' before he could say the name, all the undergraduates called out 'Scott of Girton,' and cheered tremendously, shouting her name over and over again with tremendous cheers and waving of hats." (p. 194)

The onset of Charlotte Angas Scott's progressive deafness had become apparent by the time she entered Girton in 1876 at the age of eighteen. The publicity resulting from her success was a turning point for women's higher education in England. After obtaining her doctoral degree from the University of London in 1885, Scott came to the United States to join the mathematics department at Bryn Mawr College in Pennsylvania. By 1906 she was completely deaf, and arthritis had developed so severely that she suspended scholarly publications in mathematics for several decades. During this time, however, she took up gardening seriously and even developed a new strain of chrysanthemum. Scott had many accomplishments. She was one of the organizers of the American Mathematical Society (AMS) in 1891 and coeditor of the *American Journal of Mathematics* in 1899. In 1902 and 1903 she was chief examiner in mathematics of the recently established College Entrance Examination Board. In 1894 she published *An Introductory Account of Certain Modern Ideas and Methods in Plane Analytical Geometry*. Kenschaft (1987) wrote that Scott's "impact on mathematics education in the United States was enormous" (p. 200). This was true not only in the many mathematical publications she completed on such topics as adjoint curves, equianharmonic cubics, polygons, and conics, but also in terms of her influence on mathematics education, particularly for women, in both England and the United States.

Astronomy: Women "Computers" and the Harvard College Observatory

From the earliest attempts to educate deaf pupils, astronomy had been taught—by Ponce de León, in the sixteenth century, and Abbé Sicard in the first government-sponsored school for deaf children, the National Institution for the Deaf in Paris. Sicard's student, Laurent Clerc, once described his teacher as having "directed their thoughts towards all the physical existences submitted to their view through the immensity of space, or on the globe which we inhabit; and the regularity of the march of the sun and all the celestial bodies; the constant succession of day and night; the return of the seasons; the life, the riches and the beauty of nature; made them feel that nature also had a soul, of which the power, the action, and the immensity, extend through everything existing in the universe; a soul which creates all, inspires all, and preserves all" (Waring, 1896, p. 9). Astronomy was also taught to Jean Massieu by Jean Saint-Sernin, who was responsible for the curriculum. Yet, with the exception of Goodricke in England, who died only a few years before the institution in Paris was opened, no other deaf person had pursued scientific study of the stars.

As with hearing people, the beauty of the heavens has frequently inspired deaf writers, young and old, to share their wonderments. On a shelf in a small library in London, England, sits a quaint old volume gathering dust and hardly ever read, a collection of writings of deaf children from 1845 in Rugby, Warwickshire. A young deaf woman, perhaps on a night when she had looked upward at the tranquil sky to study the celestial panorama, wrote the following words: "There are many stars whose light now glimmers for the first time before our eyes, and there may be many more whose light has not yet reached us. Is it, then, the slowness with which light travels, or the immensity of space over which it has to go, that is the occasion of this?" (Bingham, 1845, p. 117). Almost as if the young pupil's question had traveled through space and time, it was partly answered a few decades later through the extensive analysis of starlight by two American women who were adventitiously deafened, Annie Jump Cannon (1863–1941) and Henrietta Swan Leavitt (1868–1921). Both of these women were hired as "computers" at the Harvard College Observatory along with many hearing women. While they began with jobs which may be considered menial, the two deaf women rose in stature, not only providing leadership in this endeavor, but continuing a fascinating legacy begun by John Goodricke—an affinity between deafness and binary stars. In time, Annie Jump Cannon would be honored as the "Dean of Women Astronomers," receiving the Draper Medal from the National Academy of Sciences for her meticulous classification of more than 300,000 stars, and Henrietta Swan Leavitt would be considered for a Nobel Prize for formulating the law relating the period and luminosity of stars, thus opening further the field of astrophysics and providing astronomers all over the world with a yardstick for measuring the universe. Cannon lost her hearing while at Wellesley College. Henrietta Swan Leavitt was described as "extremely deaf" by the time she enrolled in Radcliffe. It is uncertain when Leavitt lost her hearing, but it was probably before she attended Oberlin College a few years prior to entering Radcliffe. Helen Keller also attended Radcliffe a few years after Cannon and Leavitt, but in response to my inquiry, Anne Enegelhart, assistant curator of manuscripts at Radcliffe, explained that there were no special programs for deaf women and that it was "sheer coincidence" that three women experiencing deafness to various degrees were there around the same time.[10]

I am uncertain if it was common for astronomers to sit in a circle during meetings, but Crowther's (1970) description of his visit to the Harvard College Observatory, during which thirty men and women took this seating formation to discuss a white dwarf Annie Jump Cannon had discovered, may suggest that an effort was made to allow Cannon to have optimal conditions for speechreading. There is scant evidence of communication between Henrietta Swan Leavitt and Annie Jump Cannon. Barbara L. Welther, historian of astronomy at the Smithsonian Astrophysical Observatory, informed me that Cannon and Leavitt worked together on the Maria Mitchell Fellowship Committee at the Harvard College Observatory, but there are few records of written communication between the two astronomers.[11] In a photograph published in an 1892 issue of *New England Magazine*, Cannon and Leavitt are seen working on stellar spectra

Annie Jump Cannon, Dean of Women Astronomers

Henrietta Swan Leavitt, American Astronomer

classification along with six other women. Also, in Cannon's talk titled "The Story of Starlight," given shortly before her death, she summarized the development of spectroscopy since the time of Newton's 1666 experiments with the glass prism and recognized Henrietta Swan Leavitt's discovery of the period-luminosity law as an important step in this history.

The Silent Angels of Mercy Hospital

In 1910, Dr. Katharine B. Richardson at the Mercy Hospital in Kansas City, Missouri, made one of the first attempts to open the doors of nursing as a profession for deaf women. Two young women, one totally deaf and the other hard-of-hearing, were chosen to enter the training program. As H. U. Andrews (1910) wrote, "Such an experiment has never before been tried in any similar institution, and the fact that it bids fair to be successful seems to point to the opening of an entirely new profession for the deaf" (p. 471).

Richardson was an attending surgeon and corresponding secretary in the hospital responsible for the admission of women to the training program. The experiment with the deaf women evolved from her interest in the deaf children of her friends. Deaf communities in several countries commended her for her efforts. In England, one writer summarized: "It is said that their work compares very favourably with the hearing nurses, and that their examination papers show a most satisfactory understanding of their subject, and the quickness with which they observe and imitate the other nurses convinces one of their general aptitude. . . . Dr. Richardson . . . is desirous that more [deaf women] should try to pass the entrance examination" (Roe, 1917, p. 214).

Richardson's interest in the problems educated deaf women faced in procuring positions commensurate with their skills and knowledge has particular historical significance, for even today the attitude about nursing careers for deaf persons presents a serious problem of discrimination. Until the time of Richardson's experiment, teaching was practically the only profession a deaf woman (as well as a hearing woman) with academic training could enter. When May Paxton, a deaf woman, approached Richardson about nursing, she agreed that the shortage of qualified personnel in many areas where the demand was high might result in the area being more accessible to deaf persons. Richardson faced obstacles at the outset, but quickly convinced the hospital's central board to give it a try. She placed no restrictions on the spoken or sign language communication skills of the deaf applicants. As Andrews wrote, "Casual objections arise at the first glance. . . . Yet, as a matter of fact, when it is explained how far the innovation has succeeded and in just what capacity the young women will be expected to serve, the most strenuous objector is bound to admit that the plan is eminently feasible" (p. 471).

The deaf applicants to the nursing program were expected to qualify on an equal basis with hearing individuals in terms of the usual requirements of health, education, and "temperamental fitness." Paxton, a graduate of the Missouri School for the Deaf, and Marion Ethel Finch, a graduate of the South Dakota School for the Deaf in Sioux Falls, and Gallaudet College, were accepted in the program. Richardson was happy with their progress, and later admitted two more deaf women, Lillie Speaker as a nurse trainee and Emma Brewington as a night helper. The nurse trainees received six dollars per month for the first three months. After they completed training, they would earn the regular rate of ten dollars per month. Brewington was in charge of linen and worked in the nursery. The others worked in the operating room during tonsillectomies and in the post-operating room as well as the nursery.

After fourteen months of training at Mercy Hospital, Finch was forced to return to South Dakota in order to keep a claim to her family's farm. She later found a position as a teacher in the Oregon School for the Deaf (1912–1917) and applied her nursing skills there. Lillie Speaker left the program when she became ill. She later married and did not return to nursing. Paxton remained for three years, but left to marry. None of the "Silent Angels" completed the nursing course.

In interviewing the deaf trainees, Andrews found that hearing nurses in the hospital were willing to write things down. Pairing hearing nurses with the deaf trainees was carefully planned, particularly for emergencies. Richardson and Miss Porter, the superintendent trainer, wrote out their lectures, something that is not necessary today with the availability of support services (notetakers, interpreters, media, etc.). The deaf women did exactly the same work as the other nursing students. Common sense called for precautions. A deaf nurse, for example, was generally not left alone in charge of a ward. However, during an epidemic of measles, Finch did manage a ward, communicating with another nurse by means of signals they had established with electric lights.

In discussing the positive effects of Dr. Richardson's program, Andrews explained that the "misfortune" of deafness for young women appeared to decrease as the promise of job security in such a profession came to them.

Other than these isolated opportunities in mathematics, astronomy, and the nursing experiment in Kansas City, Missouri, no science-related profession appeared to attract or be open to even a few deaf women during the nineteenth century or the very early decades of the twentieth. In a column titled "Deaf Women and Their Work," Hypatia Boyd encouraged deaf women to consider taking the Civil Service Commission examination to become assistant microscopists: "Deafness is no obstacle in the application for such a position, but one of the requisites for eligibility is that applicants have good, strong eyes, which can stand constant use in the manipulating of the microscope" (*SW*, *12*[10], 1900, p. 155). No deaf women followed up on her suggestion.

ELECTRICAL SCIENCE: A PROMISE OF COMMUNICATION

When Samuel F. B. Morse had the telegraph line strung from the Library of Congress to Baltimore, it passed through his business manager Amos Kendall's estate, soon to become the National Deaf-Mute College. As Gallaudet College graduates who lived near "Telegraph Hill" enjoy relating, over that line the biblical quotation "What hath God wrought!" was transmitted. But while communication through electrical transmission held great promise for improving the quality of life for deaf people, it was, in fact, a barrier wrought for them and it would be another century before any electrical device would provide deaf people direct access to telecommunications.

Much of the efficacy of telegraphed communication in the second half of the nineteenth century was dependent on the accuracy of translation of audible tones transmitted electrically over miles of lines. At first glance, one would think this was not the field of science which should be pursued by a person with a hearing loss. Yet, three influential men involved with the science of telegraphic communication were challenged by deafness through most of their lives. Somehow, they were inspired individually to contribute much to the advancement of telecommunications as a science. We might conjecture that a loss of hearing may have served as a stimulus for their work. Thomas Alva Edison (1847–1931) was nearly effervescent about how his deafness contributed to his inventive talents. But this was not at all the case for the British scientists Oliver Heaviside (1850–1925) and Sir John Ambrose Fleming (1849–1945). Strangely, none of these inventive men devoted significant time to applying their knowledge of electrical theory to helping themselves or other people with hearing losses to hear better, or to design visual or vibrotactile electrical devices to facilitate communication. Thus, for a while, the deaf inventor William E. Shaw, with only this goal in mind, struggled fruitlessly with inadequate funding to bring the telephone to the Deaf community.

Contributions to Telegraphy

Edison's work on the telegraph, telephone, and other acoustical apparatus apparently had little to do with long-standing interests in elocution or music as had been the case with Alexander Graham Bell. As Crowther (1937) has written:

> His deafness may have given him an unconscious interest in acoustical appliances, and he may have had some hope that he could invent a mechanical aid for his affliction. But it seems more probable that deafness would have created a distaste for acoustics. If that was so, Edison's mastery of his revulsion, followed by great acoustical invention, becomes psychologically still more remarkable. (pp. 383–384)

Edison was the most prolific inventor in history, credited with nearly 1300 patents, more than one hundred of them in one year (1882). Many books have been written about Edison's life and work, but the extent to which Edison attributed his hearing loss to his success, or to which his work influenced the lives of other deaf people, has not been the focus of any Edison biographies. Edison once wrote in his diary that he became deaf when he was about twelve years old:

> I had just got a job as newsboy on the Grand Trunk Railway, and it is supposed that the injury which permanently deafened me was caused by my being lifted by the ears from where I stood upon the ground into the baggage car. Earache came first, then a little deafness, and this deafness increased until at the theater I could hear only a few words now and then. . . . After the earache finally stopped I settled down into steady deafness. . . . I have been

> deaf ever since and the fact that I am getting deafer constantly, they tell me, doesn't bother me. I have been deaf enough for many years to know the worst, and my deafness has been not a handicap but a help to me. (Runes, 1948, p. 44)

This cause of Edison's deafness appears somewhat improbable, however. The deaf actor Emerson Romero (1939) argued that to damage both eardrums, the ear pulling would have to have occurred on both sides simultaneously. One of his boyhood friends, James Clancy, also recalled that Edison experienced hearing loss after a case of scarlet fever shortly after he was born (*DA*, *26*[3], 1973, p. 35). And Henry Ford, his close friend, wrote in *Edison As I Know Him* (1930) that the train incident may or may not have started Edison's deafness: "His extreme deafness dates from an operation for mastoiditis some years ago" (p. 25). Coincidentally, Edison's son, Charles, an engineer and secretary of the navy, experienced a "severe hearing handicap" after a case of typhoid fever prior to his becoming the governor of New Jersey (*Announcer*, November–December, 1948).

At least one member of the Deaf community in Edison's own time, Frank M. Howe, angrily disagreed when Edison referred to himself as "deaf." Howe preferred the term "hard-of-hearing" instead. "If he were deaf he could have had no more to do with the devising and perfecting of such an instrument as the phonograph," he wrote sarcastically, "than Helen Keller, the blind girl, could officiate as coxswain of the blimp Shenandoah" (*SW*, *37*[8], 1925, p. 402).

In response to my own query about Edison's hearing loss, Mary B. Bowling at the Edison National Historic Site in West Orange, New Jersey, explained that although nothing conclusive is available, Edison's progressive deafness apparently affected his left ear completely and, by old age, the loss in his right ear was virtually total.[12]

How was the inventor capable of receiving telegraphic signals? As Crowther (1937) explained: "This ailment made telegraph work harder, as a large part of reception at that time consisted of taking messages from a ticker by ear. There was also a benefit in deafness, as he was not distracted by the noise of instruments in other parts of the room. Perhaps the extra listening effort, which Edison had to make, increased self-control in other directions" (p. 346). Early telegraphic sounding equipment was highly mechanical, and the devices vibrated sufficiently to enable one to detect the sounds through the sense of touch. Edison's performance with the telegraph, despite his deafness, was not unique. In 1893, for example, a deaf telegrapher was mentioned in Missouri who "receives messages by putting his head against the instrument at which he is working, so that he can feel the jarring of the sounds" (*SW*, *5*[11], p. 4). In 1908, *The British Deaf News* (*5*, p. 52) described a Portland telegraph operator who was deaf. Mr. Foley worked the busiest lines by placing his forefinger on the sounder in the Western Union office and receiving the signals accurately. And, in 1916, W. C. Palmer, a deaf telegrapher, was reported as having copied train orders for 43 years through the sense of touch (*SW*, *28*[6], p. 104).

In Europe, Ernst Werner von Siemens (1816–1892), troubled with deafness since his high school days, also contributed to telegraphic science. As an officer candidate at the Prussian military and engineering school in Berlin, Siemens' left ear drum was pierced during a shooting drill. Then, a year later, when he was attempting to find a better method for igniting the charge for cannons, a pomade pot used for his experiments exploded and his right ear drum was damaged. As he explained in his memoirs: "In consequence of this mishap I have long suffered difficulty of hearing and still do suffer from time to time, whenever the closed rents in the tympana chance to open" (von Siemens, 1893, p. 31).

Siemens is well-known for his investigations of heating of the dielectric of a condenser by sudden discharge, the measurement of electrical resistance, the principle of the dynamo electric machine, and other electrical work. His work in telegraphy included improvement of the indicator telegraph of Charles Wheatstone, and his lectures on this were attended by such distinguished contemporaries as Helmholtz, Clausius, and Bois-Reymond. After developing a telegraph system using a wire with a seamless insulation of gutta-percha, he founded with Johann Georg Halske the Telegraphenbauanstalt von Siemens and Halske, which contracted to build a telegraph network in northern Germany and to lay underwater cables for several countries. Siemens helped to design the *Faraday*, the first special cable-laying ship, which laid five Atlantic cables in ten years. He was made a knight of the Prussian Order, an honor conferred for his services to science and industry. Emperor Frederick III conferred nobility upon him in 1888.

John Ambrose Fleming, a professor of physics and mathematics at University College at Nottingham, resigned to begin consulting work with a branch of one of Edison's companies in London. Under Edison and, later, Marconi, Fleming was required to sign a contract agreeing that any inventions belonged to them. Eventually Fleming felt the financial rewards were inadequate, and stopped working for them. Fleming is best known for his invention of the vacuum tube, which he preferred to call the "electric valve."

Sometimes Fleming's deafness gave rise to humorous situations. W. H. Eccles (1945) described one occurrence which happened while Fleming was teaching:

> This deafness, by the way, was utilized by his students when in jubilant mood; on such occasions a stranger passing the door of the lecture room might hear a sudden clamour and its sudden cessation, and if he lingered he would hear the uproar switched on and off at intervals. Inside the lecture room he would have seen that when the professor turned to the blackboard there was pandemonium, and when he faced his class there was silence. The timing had to be lively as Fleming was quick in his movements. (p. 663)

Another incident happened when Fleming lectured at the Royal Society of Arts. He was being introduced by Sir Henry Truman Wood, who was not visible from the lecture table, and Fleming, assuming that Sir Henry had finished his introduction, began his lecture. The two of them continued for some time, but Fleming soon had the audience to himself.

When Fleming presented on the topic "Electric Resonance and Wireless Telegraphy" in the Royal Institution lecture theater on June 4, 1903, arrangements had been made to receive a wireless message from the Marconi Station at Chelmsford during the lecture. Arthur Blok, Fleming's friend, related the story as follows:

> One of the Marconi Company's staff was waiting at the Morse printer and while I busied myself with demonstrating the various experiments I heard an orderly ticking in the arc lamp of the noble brass projection lantern which used to dominate this theatre like a brazen lighthouse. It was clear that signals were being picked up by the arc and we assumed that the men at Chelmsford were doing some last-minute tuning-up. But when I plainly heard the astounding word "rats" spelt out in Morse the matter took on a new aspect. And when this irrelevant word was repeated, suspicion gave place to fear. (MacGregor-Morris, 1954, p. 128)

Following the "rats" message was a limerick about a young lady from Italy, and then some quotations from Shakespeare. Blok continued:

> Evidently something had gone wrong. Was it a practical joke or were they drunk at Chelmsford? Or was it even scientific sabotage? Fleming's deafness kept him in merciful oblivion and he calmly lectured on and on. . . . All seemed well—a testimony to the spell of Fleming's lecturing—until my harrassed [sic] eye encountered a face of supernatural innocence and then the mystery was solved. The face was that of a man whom I knew to be associated with the late Mr. Nevil Maskelyne in some of his scientific work. (p. 128)

Maskelyne later assumed responsibility, admitting to having sent out the messages to show that the Marconi system of signaling was not foolproof against interference.

Fleming's extensive experiments on transmitters, receivers, wireless telegraphy, and the electric valve earned him many honors. In addition to being a fellow of the Royal Society, he was the recipient of the Society's Hughes Medal and the Faraday Medal and he was awarded the Gold Medal of Honor of the Institute of Radio Engineers.

The Telephone

It is not surprising that with so many clever minds working on the problems of transmitting multiple electrical signals with audible dots and dashes in the nineteenth century some began to consider the process of transmitting more complex sounds, including human speech. "Speaking machines" or "talking machines" had long been an interest among scientists and inventors. Christopher Wren built one in the seventeenth century. Erasmus Darwin, the grandfather of Charles Darwin, was hard at work on a "speaking machine" in 1771, and Robert Willis, a fellow of the Royal Society, designed another in 1829, perhaps the first systematic examination of the production of vowel sounds before the advent of X-rays (Hodgson, 1953). In 1843, a Mr. Reale exhibited before the American Philosophical Society a speaking machine which produced various letters and words. Frustrated at its failure to work properly, the inventor destroyed the instrument, which had taken him sixteen years to build (Timbs, 1860).

In 1861 Phillip Reis in Germany developed an electrical device he called the "telephone" that produced speech by vibrating a diaphragm with the sound energy from the human voice. Reis's "telephone" was not practical because it required frequent adjustments. Its principles of operation were known by Edison, and Joseph Henry had also explained them to Alexander Graham Bell (Bruce, 1973). The word "telephone" is actually much older than Reis's invention, however. Scientists in the eighteenth century used it in reference to a speaking tube or megaphone. Claude Chappé called the "mouth trumpets" which he had substituted for the optical telegraphy system in France "telephones." Huth's use of "telephone" in a paper he wrote in Germany on acoustic instruments was much closer to the modern meaning of the word.

Alexander Graham Bell's first love was teaching deaf students. While he had no hearing impairment, his mother, Eliza Bell, was deaf. She could not speechread well, and Alexander would fingerspell in his conversations with her. Later, Bell fell in love with and married a young deaf woman, Mabel Hubbard, the daughter of Gardiner Greene Hubbard, who had assisted him in financing his scientific research. Bell's interest in the telephone diaphragm was sparked by young Mabel's comment that she could feel sounds through her muff when walking in the winter weather. They married on July 11, 1877, a year after Bell demonstrated the telephone at America's centennial celebration in Philadelphia. Bell gave his deaf bride all his financial interests in the telephone as a wedding gift.

Bell had a long-standing interest in the electrical transmission of messages. One of his patents was based on the principle of converting a vibration into a special form of a vibratory circuit breaker. Another was the autograph telegraph, which could copy handwriting at a rate of about 3000 letters per minute. It was a combination of his interest in the multiple (or "harmonic") telegraph and in making speech more visible for his wife, which led him to the discovery for which he is now most famous (Burlingame, 1964).

Bell found that he could induce an electric current in a coil of wire wound around a permanent magnet by vibrating an iron diaphragm near the magnet. If a voice were made to vibrate the diaphragm, the amount of current induced in the coil would vary. Then, the reverse process could also be made to happen. A current sent to a similar device would vibrate the diaphragm and reproduce the original sounds. One difficulty he had was in acquiring sufficient energy to move the diaphragm. The current induced by a human voice was faint. As the story is told, Bell was working in his laboratory on March 10, 1876, when he accidentally spilled some sulfuric acid used with the telephone apparatus. He called for his assistant: "Mr. Watson—come here, I want you." Thomas Watson was in a different part of the building and heard Bell's voice—not through the walls—but on the device. It was entirely by accident that Watson had attached the reed too tightly against the pole of an electromagnet and that the right man had the device to his ear at the right time (Burlingame, 1964, p. 74).

The telephone was exhibited at the Centennial Exposition in 1876. Here it was nearly ignored until it attracted the attention of some foreign visitors, among them Dom Pedro, emperor of Brazil, and Sir William Thomson, afterward Lord Kelvin. Sir William and Lady Thomson were reportedly so fascinated by Bell's device that they changed places repeatedly, speaking and listening, and enjoying the invention thoroughly.

On October 31, 1877, Alexander Graham Bell honored the members of the Society of Telegraph Engineers in England with a "delightful and inspirational presentation" (Appleyard, 1939, p. 57). Appleyard described the special moment in history when W. H. Preece exhibited the Bell telephone before a very large audience in England:

> Many learned men were present. There is one remarkable feature of a learned meeting: when you call upon a learned member to make a learned remark he frequently makes a foolish one. Now, I selected perhaps one of the leading scientific men of the day, and I placed the telephone in his hand. It was in connection with a similar instrument 55 miles away. Of course we expected to hear from him some learned axiom, some sage aphorism, or some wonderful statement: but after hesitation he said into the telephone "Hey diddle diddle—follow up that." He then rapidly put the telephone to his ear, and reported with much glee, "He says: Cat and the Fiddle." My assistant, who had been at the distant receiver during this demonstration, was asked by me the next day if he understood "Hey diddle diddle." He said "No." "What did you say?" My assistant replied, "I asked him to *repeat*." (p. 57)

Bell's telephone was considered an excellent receiver, but it was not an effective transmitter. The problem of faint currents remained. Edison came to the rescue with a carbon transmitter, consisting of a vibrating plate abutting against a carbon button, which greatly improved the transmission capabilities of the telephone a year later. Vibrations from human speech varied the pressure on the disc and this, in turn, caused the resistance of the carbon and the electric current to vary.

In a familiar refrain, Edison claimed that his deafness was responsible for the perfection of the carbon transmitter. According to him, he was unable to hear anything on Bell's telephone. "The telephone as we now know it," Edison explained, "might have been delayed if a deaf electrician had not undertaken the job of making it a practical thing" (*SW*, *38*[4], 1926, p. 178). Edison coined the call word "hello," which quickly spread in use. He invented a new nonmagnetic type of telephone receiver, one that did not conflict with Bell's patent. The electromotograph receiver was based on the principle of electrolytically varied friction between conducting surfaces in contact.

Fleming's British contemporary, Oliver Heaviside, was an eccentric hermit who greatly advanced telephone transmission theory after Bell's invention came into use. Heaviside developed the operational calculus and was one of the first to suggest that the Earth's atmosphere could make long-distance transmission of electromagnetic waves possible. On Heaviside's hearing loss, Nahin (1988) summarized:

> Heaviside's deafness (due to a bout with scarlet fever) was to bother him all through life. Even in his published work he referred to it; for example, in one article when discussing how the human ear can decipher a voice in the midst of telephone circuit noise: ". . . some remarkable examples (of interpreting indistinct indication) may be found amongst partially deaf persons who seem to hear very well even when all they have to go by (which practice makes sufficient) is as like articulate speech as a man's shadow is like the man." (p. 15)

Oliver Heaviside, British Electrical Scientist

In a letter to J. S. Highfield, president of the Institute of Electrical Engineers, Heaviside complained, "My eyes are very bad. That is nothing new. It is connected with my scarlet fever ears" (Nahin, 1988, p. 291). Throughout life his hearing appeared to fluctuate. He enjoyed music, but in one of his writings on his attempts to enjoy it, he described himself as "very deaf" (Appleyard, 1939, p. 220). During rare moments, his hearing would improve, but in one of his private papers was found a handwritten note indicating his displeasure with the difficulties he was experiencing: "Episode in the struggle for life. Got rid of deafness partly. . . . Everything in this life that you want comes too late" (Appleyard, 1939, p. 217).

Heaviside seldom left his house. A friendly policeman would bring him food, blowing his whistle through the letter-box to attract his attention. If Heaviside failed to hear the whistle, the policeman would leave the food outside, and there it would remain until the secluded scientist was disposed to take it. He was a quarrelsome individual, and his life was a series of disappointments and difficulties. The extent to which his introversion was brought on by his deafness is unknown. Heaviside appears to have disdained social contacts in general. He never attended the meetings of the Royal Society, to which he was elected a fellow in 1891.

The Microphone and Phonograph

Edison's improvement of the microphone was another important advance; it followed his work on the telephone transmitter and was reported worldwide. In England, Preece's lecture on Edison's device was described in the May 30, 1878, issue of *The London Times* and induced the following letter from an excited deaf individual:

> Will you allow me, through your columns, to appeal to scientific men in behalf of a large and unfortunate class of individuals, young and old, in Her Majesty's dominions . . . who are deaf? Their case is most pitiable. Cut off from all human intercourse by the voice, theirs is, indeed, a sad and fearful lot. Science hitherto has done little or nothing for them, either in curing or in assisting hearing by instruments. The ordinary ear trumpet is a poor, useless thing; the lengthened tube is a better aid, but very unportable and unsightly, and does not assist in hearing general conversation in the room. Can it be that the microphone is likely to give the much-desired assistance which the deaf have so longed for, but longed in vain? If only, Sir, you could and would by the insertion of this letter induce men of science to turn their attention to this subject, and to succeed in making the deaf hear by the microphone, you would earn the deepest gratitude of thousands who are so grievously afflicted.
>
> I am, Sir, yours truly,
> ONE WHO IS DEAF

In the June 1 issue of the *The London Times*, W. B. Dalby, an aural surgeon at St. George's Hospital, agreed with the deaf correspondent about the lack of technical advances to help those with residual hearing, but he took issue with the comment about medical treatment, detailing a number of breakthroughs.

He explained that the microphone as an aid to hearing was then under investigation. But, in a practical sense, Dalby's comment was premature. Deaf people would have to wait another forty years for the vacuum tube hearing aid, and thirty years beyond that for one transistorized and compact. And as George Wherry's article published in *Nature* (December 12, 1901) indicated, the use of the horns from bighorn wild sheep for ear trumpets continued to be valued.

The microphone was an important component of later devices which amplified and measured hearing loss. A study of Edison's writings indicates that he had sufficient residual hearing that he would have been greatly helped by amplification. But as with some individuals today, Edison was clearly not interested in hearing aids. Nor was he interested in applying his own inventive talents to the development of such. Physicians called upon him to submit himself to treatment, but he refused. He was frequently troubled with severe earaches at Menlo Park, but his love for science and invention outweighed his desire to hear. He once wrote to a deaf correspondent who had queried him about a relief for deafness and explained that for a period of time he had conducted experiments with an assistive device to improve hearing, but that before the investigation was completed, it had to be abandoned in favor of other important work. Edison believed his idea might prove a boon to some people with deafness. One might conjecture that, with his interest in profit, he might have been pursuaded to continue had he seen a potential market. In 1893, Baron Leon de Lenval of Nice offered a prize of 3000 francs to the inventor of a "microphone apparatus for improving the hearing of deaf persons" (*SW*, *5*[11], p. 9), but this did not seem to attract much interest. At least one periodical published for the Deaf community in England warned its readers in 1895 not to "indulge unduly in the hopes of an invention which will ameliorate their sad condition, but at the same time a message of this sort from the man who invented the phonograph and the microphone cannot fail to be of tremendous interest to the deaf."[13]

Another invention Edison attributed to his hearing impairment was the phonograph. "Deafness, pure and simple," he said, "was responsible for the experimentation which perfected the machine. It took me twenty years to make a perfect record of piano music because it is full of overtones. I now can do it—just because I am deaf" (Runes, 1948, p. 54). Earlier Edison had invented an automatic recording telegraph. As a paper disk rotated on a vertical axis the embossed marks from incoming messages caused a lever to lift up and down. When the lever was made to vibrate rapidly, the sound was raised to a musical pitch. The next logical question was—why not speech? From Edison's involvement with the improvement of the telephone, he knew that a diaphragm could be made to vibrate with speech sounds. When he covered a grooved cylinder with tin foil and connected a diaphragm with a needle, he found that speech caused the vibrations to emboss the tinfoil as the cylinder rotated.

During my visit to Edison's laboratory in 1986, I had Mr. Howe's skepticism in mind (see "Contributions to Telegraphy," above) when I asked about how Edison could depend on audible sounds alone to perfect the phonograph. A staff member showed me the wooden casing on Edison's earliest phonograph. Edison, he told me, could not hear the phonograph two feet away from it. Tooth marks indicated where Edison's jaw had firmly clamped onto the cabinetry. By placing his teeth on the wooden frame, Edison was able to pick up some vibrations of the music as they traveled through his jaw to his inner ear.

Electrical Communication: Boon and Nemesis

Indirectly, Bell's invention of the telephone and Edison's subsequent work on related devices soon led to other discoveries which were of benefit to deaf people. Newspapers and journals over the decades following the Centennial Exposition frequently reported on the applications of the technological advances to aid people with deafness. One could find in 1895, for example, "trumpet mouths" connecting the telephone with pews occupied by wealthy deaf parishioners in a leading London church. In 1896, Dr. Bertram Thornton reported in *The Pittsburg Dispatch* on his efforts to use the telephone for instructing a group of deaf children. The primary advantage was that the children were able to see the lips while the teacher was speaking. This was not possible with the "speaking trumpet."[14] The invention of the telephone also opened up some new employment opportunities. Roe (1917), for example, reported that 150 deaf men were hired by a Chicago factory to assemble the delicate mechanism of the telephone; the employer considered this "a special kind of work in which they can excel" (p. 136).

But, in effect, the telephone was one of the first major technological breakthroughs that placed deaf people at a disadvantage, in this case persisting for nearly a century.

Although the interest of Edison, Heaviside, and Fleming in telecommunications was both theoretical and pragmatic, it was not until the 1960s that deaf people would generally benefit from the technology derived from their work. On the other hand, as I will discuss later, there is no doubt that deafness was the driving force behind the physicist Robert H. Weitbrecht's work in the 1960s in developing a device to enable deaf people to use the regular telephone along with a teletypewriter keyboard, thus finally bringing Bell's wonderful invention to the world of people who could not hear. The American Deaf community claims Weitbrecht, and rightly so, as a hero. In any event, a compilation of the most significant inventions which have changed the lives of deaf people would surely rank Weitbrecht's telephone modem near the top. Yet, the invention of this device may have been much delayed had it not been for earlier work by Edison, Fleming, and Heaviside. Each of these men, their interest piqued by Alexander Graham Bell's telephone, dismissed his deafness as an obstacle to pursuing ideas for perfecting the telecommunication science of the time.

EMERGING PROFESSIONS

Invention

It is more than conjecture to assume that there was a connection between the inventiveness of Americans in the nineteenth century and their independence of behavior. As Crowther (1937) has described metaphorically, the colonies attracted people through a kind of "natural selection" from the more independent of the Europeans, and he suggested that the immigrants to America were slightly more inventive in their abilities. Following the Civil War, the population of the United States increased by millions and large-scale manufacturing grew rapidly. With these changes the development of labor-saving devices was welcomed by industry. Along with Edison, many other deaf men pursued invention as a profitable venture.[15] But Edison, unlike most other deaf inventors, frequently publicized his belief that there was an inseparable bond between deafness and invention and that intensive thought and concentration were easier to achieve when one is deaf. In the sense that deaf people have always had to be inventive to deal with the attitudinal, communicative, and other barriers associated with living with a hearing loss, Edison's notion holds truth. And in the technological sense, the gift of inventive talent among deaf people during the last half of the nineteenth century did become prominent enough that it attracted the attention of patent attorneys, who ran advertisements in various newspapers published for the Deaf community.

The patents awarded deaf people in this period were many and varied. Fisher A. Spofford and Matthew G. Ruffington in Columbus, Ohio, were issued a patent for their water-velocipede. Anton Schroeder, a graduate of the Minnesota School for the Deaf and St. John's University, turned to invention and acquired a number of patents in the building hardware line, some associated with Stanley Works in Connecticut. In 1897 Schroeder won an award at the state fair of Minnesota for his hanger for screens and storm windows. He also invented an adjustable candle lighter that was widely used in churches, convents, academies, and hospitals, and his streetcar curtains were used in St. Paul. Other inventions during this period included Philip A. Emery's sundial and sextant, E. M. Jacobs' "Arrow" razor blade honers, and Dale Paden's pipe wrench and vise-grip pliers. In 1881 Henry Haight's incubator for hatching chicks won him a gold medal at the National Poultry Exhibition. In an 1871 issue of *Scientific American*, Paxton Pollard's toy cart was predicted to become "more popular than the jack-in-the-box." A decade later John R. Gregg, whose deafness resulted when a headmaster angrily knocked his head against a classmate's, organized a publishing company and patented his Gregg Shorthand method. He became a millionaire with his development of shorthand and was awarded an honorary Doctor of Commercial Science degree by Boston University. Many automobile and bicycle inventions were also reported in the literature of the Deaf community. James P. Burbank, one of Alexander Graham Bell's pupils, established the first magazine in America devoted to the subject of bicycling. Robert Carr Wall, a manufacturer in Philadelphia who had attended the Western Pennsylvania School for the Deaf, built the first "safe" bicycle in the city, improving on the English version by having two wheels of the same size. Wall also manufactured engines. In 1900, he built and sold Philadelphia's first gasoline automobile for $1800, and Packard Automobile Company asked him to design and build a rattle-proof windshield. John Russell Francis, a lens maker at Eastman Kodak Company in Rochester, New York, received patents for an automatic hasp lock and a combined automobile windshield blind and sunshade.

Flying also attracted interest. Edison designed a machine mounted on wheels, whose wings were moved by an engine. Publisher James Gordon Bennett gave him $1000 to further experiment with flying; Edison then built a helicopter which did not fly because the metal and motors were too heavy. Only a few years after the Wright brothers' historic flight in 1903, several deaf inventors were reported to have flown machines of their own. In Pittsburgh, Pennsylvania, on July 8, 1909, many people watched as Walter Asbury Zelch, twenty-three years old, flew a bicycle-propelled "aeroplane" across the valley between Mt. Washington and Duquesne Heights. Zelch built five-foot-wide canvas wings which spanned twenty-two feet across. One revolution of the two pedals caused the propeller to rotate four times, and the aircraft was steered by a handlebar. According to the July 1909 issue of the *Philadelphia Record*, the airplane glided across the valley without mishap, but when Zelch became excited, he turned the handlebar in the wrong direction and broke the bevel wheel in the subsequent collision with a tree (*Maryland Bulletin*, February 23, 1910).

Some deaf inventors turned their skills to improving the quality of life for deaf people. Walter Lazell's novel alarm clock dropped a light oak beam on deaf sleepers. In retrospect, this invention appears humorous, but many elderly deaf people have related to me similar alarm clock improvisations of their own. William E. Shaw designed a device which projected an image of the operator's mouth and lips upon a screen for the purpose of facilitating speechreading. In terms of direct benefits to deaf people, in particular, Shaw was one of the most productive deaf inventors of the nineteenth century. He devoted much of his life to assisting the Deaf community. Shaw attended the Portland School for the Deaf in Maine and then the American School for the Deaf in Hartford, Connecticut, where he graduated. He was credited with at least fifty inventions specifically benefiting deaf people. Many of the inventions were visual and vibratory signalers such as those used to help deaf people answer the door. Shaw's "Bell-less" doorbell was very popular, as was his novel alarm clock (the "Pillow Shake") which was advertised in the *Boston Sunday Herald* and *The Silent Worker*. He traveled around the country, throwing "electrical parties" for forty or fifty deaf men and women at a time in order to demonstrate his devices. He knew the demand for these inventions would remain low and he never expected to become wealthy from them. The complexity of Shaw's various inventions is both impressive and humorous. His "Baby-Cry" signaler to wake up deaf parents included a croquet ball attached to a spring which was tied to the sleeping baby. Whenever the baby swung its arms if awakened by a noise or if sick, the croquet ball was released to hammer against the parents' bed frame. Shaw's alarm clock was a marvelous contraption which not only told time but alarmed the sleeper by agitating a lever which vibrated the pillows on the bed. It also ignited a match which then lit an oil lamp, and the light from the lamp awakened the deaf sleeper if the vibrating pillow failed to accomplish that purpose. Further, the device closed a circuit which sent an electric current to an incandescent lamp in front of a parabolic mirror, reflecting the rays onto the face of the person asleep. A string connected to a hammer also fell upon a dynamite cap, making a loud explosion, and the entrance of burglars would also give notice to the deaf person by any of the above methods by means of wires connected to the doors and windows. Finally, electric thermostats placed around the premises were connected to the device to signal a fire. Any combination of these methods was possible, and as an anonymous writer in the *Boston Sunday Herald* said on March 8, 1903, the invention would "awaken the sleepiest deaf man that ever lived."

Only a decade after the Centennial Exhibition in Philadelphia, in which Bell's telephone was demonstrated, Willie Mosher, a young deaf man, proposed a "Mute's Telephone," whereby a key moved on one end of the telephone line would move a similar one on the other end (*SW*, *8*[5], 1896, p. 5). Sixteen years later, William E. Shaw designed and built a "Talkless Telephone," a light board with letters and figures attached to a battery. Each deaf person on the telephone had a keyboard connected to the rows of flashing lights and transmitted and received messages, letter by letter (*Daily Evening Item*, March 12, 1912). It was a precursor of the modern digital electronic devices. Shaw's invention enabled his friends to use Bell's telephone, and Bell himself was so impressed that he wrote the deaf inventor a letter which Shaw cherished until the day he died. "Congratulations on your splendid work," Bell wrote, "which now brings happiness and the human voice to those who could not otherwise enjoy them" (*The Cavalier*, *5*, 1944). Edison also heard of Shaw's work and Shaw joined Edison's laboratory in West Orange, New Jersey, as an electrician. Three other deaf employees worked there with him. In the same year John H. Norris, a deaf electrical engineer in Atlanta, also designed a telephone for "his fellow unfortunates." Norris's telephone was, in reality, a telegraph system which made use of lights: "At each station is a board upon which are the letters of the alphabet and numbered push buttons." A cable of seventy-two wires made the invention impracticable (*SW*, *25*[1], 1912, p. 13). The "talkless telephones," as they were designed in this period of history, were far too complicated and never gained widespread use.

In Europe, several deaf inventors also worked toward making life more comfortable for deaf people. Hieronymous Lorm (Heinrich Landesmann), deafened at sixteen, and completely blind later, invented a touch alphabet for deaf-blind people to use in Germany. After his death in 1902, his daughter Maria published a pamphlet which helped to spread the alphabet's use throughout the world. And a physician named Max Albert Legrand of Paris, described as having "bravely faced his affliction, and, since the day it befell him, [having] tried to do the other deaf as much good as possible," opened a bureau of "mutual help and information" and invented a dactylophone consisting of forty-two letters manipulated by keys which when pressed stood up and helped his deaf friends to communicate with others (*SW*, *25*[6], 1913, p. 103).

Meanwhile, on the humorous side, back in America J. C. Chester, a deaf man from Montana and a graduate of the Iowa School for the Deaf, was reported journeying to Washington, D.C., in a covered wagon with his wife and three small children. He purportedly invented an instrument that would enable deaf people to hear and he was on his way to have it patented. The last report indicated he made it as far as Syracuse, New York (*SW*, *21*[2], 1908, p. 33).

Engineering

Another viable occupation for deaf people which began to emerge at the turn of the century was engineering. In 1895 James L. Smith of Minnesota developed a list of 250 occupations pursued by deaf persons (Porter, 1896). Civil engineering was not on this list, and it might be inferred that there were no deaf engineers. But Ormond E. Lewis was then in the employ of the Purdy and Henderson firm, and other deaf men had succeeded in engineering as well. Lewis, who lost his hearing at the age of two years, studied at the Clarke Institution at Northampton, Massachusetts, and later entered the Kendall School in Washington, D.C., from which he graduated. After a period with his family business as a bookkeeper, he enrolled in the Kansas City School of Design and studied in the office of McKim, Mead and White, a Kansas City architectural firm. At Purdy and Henderson he worked with other civil engineers whose speciality was in the construction of steel buildings and bridges.

Melvin Hoyt Wheeler, also a graduate of the Clarke Institution, completed a degree at Harvard in 1902, along with his deaf brother Homer and Robert Pollak, all in civil engineering. Tilston Chickering, also deaf, earned a degree from Harvard in mechanical engineering that same year. Melvin Wheeler was first employed by the American Bridge Company in Pittsburgh, Pennsylvania, then later by the Berlin Construction Company in Connecticut. In 1907, he designed several large office buildings and highway bridges in Los Angeles.

T. Kennard Thomson lost his hearing at about two years of age and was, according to one writer, "greatly handicapped in school and college because of it" (Holt, 1925, p. 675). He graduated from the University of Toronto at the head of his class. In 1886, Thomson received a bachelor's degree in civil engineering and later earned a doctor of science degree. He became chief engineer for a firm of contractors and was involved in the construction of skyscrapers. He completed 129 bridges on one railroad alone, and served on the Board of Engineers of the New York State Barge Canal. Thomson patented many inventions and published and presented numerous papers with the help of an "earphone." At least fifty of his papers were published by the Society of Civil Engineers. Thomson proposed a "utopian city" to alleviate congested conditions in the lower part of Manhattan. The elaborate and detailed design included the building of a double seawall and pumping out the water to reclaim nine square miles of land, and a comprehensive plan providing for buildings covering whole blocks, with rapid transit, pollution-free hydroelectric power, gardens, playgrounds, and museums, and with four levels of transportation in each building, which would house ten to twenty thousand people as a community in itself with a post office, fire department, gymnasium, and medical services. Thomson was described as an engineer "whose mental vision is as keen as the physical vision which supplements his deafened ears" (Holt, 1925, p. 675).

Another civil engineer was John H. Clark. Born in the small town of Panguitch, Utah, on May 17, 1880, he was profoundly deafened by spinal meningitis at the age of ten. After recovering from a coma, he entered the Utah State School for the Deaf at Ogden, where his teachers sensed his abilities and encouraged him to prepare for Gallaudet College. He graduated from the Utah School and entered Gallaudet in 1897. Tutored by Percival Hall (who later became Gallaudet's second president), Clark excelled in mathematics and English and in his senior year was elected editor of the student magazine, *The Buff and Blue*. In 1902 he graduated from Gallaudet College, returning to Utah, where he took up further studies in civil engineering, and in 1904 he began working for the Forest Service, surveying the Grand Canyon. Clark was cited by President Theodore Roosevelt for his interest in the conservation of natural resources (Smaltz, 1949).

One of the first deaf electrical engineers in the United States was Harris Joseph Ryan (1866–1934). Ryan gradually lost his hearing over a period of fifteen years and used a "flexible conversation tube" and other inventions of his own to assist him in communicating. With these devices he was able to teach classes at Stanford University. As Holt (1927) explained: "He has always accepted the condition in a matter of fact way, cheerfully and philosophically, never allowing it to cloud his outlook on life or change his attitude toward his work" (p. 211). Ryan graduated from Cornell University in 1887. He spent a year with an electrical firm in Lincoln, Nebraska, then returned to Cornell to teach until 1905. He was elected president of the American Institute of Electrical Engineers in 1923 and was awarded the Edison Medal in 1927 for "meritorious achievement in electrical science or the electrical arts." At the formal opening of the Ryan High Tension Laboratory, Ryan generated over two million volts in an elaborate electric arc display to thrill spectators. He and his research team spent years investigating the transmission of high-voltage electrical energy with minimal loss for the state of California. Durand (1938) explained in his National Academy of Sciences memoir that following Ryan's retirement in 1931, he occupied himself with "studies relating to the deaf and the application of electrical aids to hearing for the alleviation of this all too common handicap to human activity" (pp. 290–291). Ryan submitted himself as a subject in experiments with later devices developed in the research laboratories of the American Telephone and Telegraph Company and was enthusiastic in his efforts to help others benefit from these studies.

Ryan's chief work focused on the improvement of long-distance transmission of power, and he earned many honors for his investigations. Evidence of his influence as an engineer and "faculty man" (which he enjoyed calling himself) came in many forms, but one letter received by his wife following his death indicates the international importance of Ryan's work. As Mr. S. Motomura related, the Stanford Alumni Association of Japan held a memorial dinner at which many shared their reminiscences of their days at Stanford "in which the greatness of Dr. Ryan as both scientist and man was brought out. . . . after all had silently bowed before the photographic likeness of Dr. Ryan, as a last tribute, the meeting adjourned" (Durand, 1938, p. 299).

George Bryan Shanklin (1870–1947) was also prominent in electrical science during the first half of the twentieth century, contributing numerous patents in the area of electrical power and high-voltage cables. He was eight years old when he was deafened from scarlet fever. He first attempted to learn through private tutoring, then entered the Kentucky School for the Deaf. After two years at Gallaudet (1905–1906), he left to pursue a degree in electrical engineering at the University of Kentucky, from which he graduated in 1911. His professors frequently gave him written copies of their lectures. Shanklin's grades were so impressive that he was recruited for employment before he graduated. He began working in experimental testing for the General Electric Company in Schnectady, New York, where he pursued additional training in the operations of electrical apparatus. He was such an asset that the company sent him to Europe on many occasions to study the state of the art in electrical appliances. Because of Shanklin's deafness, his supervisor at first had misgivings about receiving satisfactory work. Shanklin not only did as well as his hearing coworkers, but he soon became head of testing, engaging himself in research and development work on high-voltage insulation designs. He was a consulting engineer in charge of a laboratory under the eminent Charles P. Steinmetz.

Gallaudet College followed Shanklin's progress as he experienced increasing success. As the editor of Gallaudet's alumni bulletin lamented, it was regrettable "that the hazers of those days did not treat Bryan so as to secure his loyalty to Gallaudet" (*B&B*, *25*[3], 1916, p. 82). Shanklin was held up as a model for others, "for in texture and breed he is the type of student that Gallaudet greatly needs."

For twenty-six years, Shanklin was the chief engineer responsible for the power cable division in the central station department, and then became manager of commercial engineering until his retirement. In this position, he provided consultation to the hydroelectric commission in Quebec, the Canada Wire and Cable Company, and the Potomac Power Company. Shanklin's work with the development of oil- and gas-filled high-voltage cables won him many honors. He was awarded seventeen patents and twice received the Coffin Award, General Electric Company's most prestigious award for employees. In 1939, Shanklin received a citation from the Engineering Societies of New England. He was also a fellow member for life in the American Institute of Electrical Engineers and is credited with the discovery of ionization in high-voltage cables. From this discovery, he developed new ways to control the ionization, engineering most of the oil-filled cable installations in the United States. He published many papers on electric cables and was very involved in electrical engineering organizations.

Dentistry

Another field of study which opened up for people who were deaf in the nineteenth century was dentistry. John Kitto had been trained in the late 1820s through an apprenticeship with A. N. Groves in Exeter, England. The elder son of a Plymouth mason, he lost all of his hearing in a fall from a ladder when he was a youth. While living with Groves' family during the apprenticeship, he spent four to five hours a day developing skills in dentistry. "The business of a dentist is of a very various nature," he wrote to a friend, "*The employment is agreeable* . . . and my part of it, as yet, consists in the formation of artificial teeth, not from ivory, but a harder substance, the tusks of certain foreign animals" (Thayer, 1862, p. 224). Kitto turned to missionary work and writing.

As the nineteenth century came to a close, three deaf dentists, whose names were not provided in the *Report of the Fourth World Congress of the Deaf* (1893) were practicing in England. Two others, also unidentified, were mentioned in a report from Queensland, Australia in 1910 (*SW*, *22*[8], p. 156). It is not known how they received training, but they were probably "mechanical dentists," like John Kitto, responsible for constructing and applying artificial teeth, plates, and other appliances to correct for irregularities. In 1911, it was reported that a "considerable number of the deaf are said to be taking up the occupation" (*VR*, *12*[10], p. 651).

In 1893, the editor of the *Mt. Airy World* at a school for deaf pupils in Philadelphia (now the Pennsylvania School for the Deaf) suggested that dentistry might be a promising field for deaf persons. Over the next few years a number of editors of other publications in the Deaf community responded. "The [Mt. Airy] *World* has evidently never heard of Ohio," wrote the editor of the *Ohio Chronicle*. "One of our own graduates, George Evans, of Springfield, Ohio, took up dentistry and practiced it successfully for several years until he went into the business of manufacturing agricultural implements" (*SW*, *10*[9], 1898, p. 133). The editor of *The Western Pennsylvanian* recognized Dr. William Hawk, a graduate of the Indiana School for the Deaf, whose business was located in Pittsburgh. Several other deaf dentists practiced around the turn of the century. Bernhard Ernest Ringnell, a graduate of Gallaudet College and the University of Minnesota School of Dentistry, offered his services in Iowa. He was born in Kalmar, Sweden, and became profoundly deaf in infancy from scarlet fever. E. C. Weinrich was employed in Chicago at the J. L. Dunkley Dental Laboratory Company. He was educated at the Lutheran School for the Deaf in Detroit, Michigan, and studied at the American Dental School of Chicago. Arthur H. Clancey, a graduate of the Ohio Dental College in 1895, took up his father's practice and was a member of the Cincinnati Dental Society and the Ohio State Dental Society. Clancey was also involved in the National Association of the Deaf. He employed a hearing assistant while serving a "large and fashionable clientele."

George K. Andree was probably the first deaf person to attain eminence in the dental profession. He was deafened at the age of twelve and attended the Michigan School for the Deaf and received his bachelor's degree from Gallaudet College in 1902. After this, he entered the Georgia School of Technology, where he began study in civil engineering and coached the football team for two years, then the University of Michigan School of Dentistry, where he received his doctorate. He was appointed by Governor Murray to serve on the Board of Dental Examiners of Oklahoma, and was elected president of the Oklahoma Dental Association. Andree was also a member of the legislation committee of the National Dental Association.

The deaf dentist most often mentioned during this period was Edwin W. Nies, perhaps because he was also an ordained minister in the Deaf community. Nies, deafened by spinal meningitis, graduated from the Lexington School for the Deaf in New York City in 1905 and from Gallaudet College in 1911. When the University of Pennsylvania accepted him, officials were not aware of his deafness, however, and upon learning that his bachelor's degree was from a college for deaf people, they withdrew the invitation (Garretson, 1950). Determined to get into dental school, Nies took the letter of rejection to Mary Wallace Weir, the dean of a private school for young women. Weir contacted the dean of the dental school, Dr. Edward Kirk, and made appointments.

Fortunately for Nies, Kirk had himself been the attending dentist at the Pennsylvania School for the Deaf during the earlier years of his practice and saw no reason why a deaf person could not become a dentist. The University of Pennsylvania subsequently admitted Nies to a probationary program, in which he joined a class of 150 dental students. Nies's professors offered him assistance, and, in exchange for lecture notes, Nies offered to his classmates his skills he had obtained from his experience in his uncle's laboratory.

Nies graduated in 1914, but could not find a job as a dentist. He took a position in a dental laboratory, at first not revealing his credentials until his coworkers had become accustomed to his deafness. After a year in dentistry there, Nies set up what became a successful private practice in Washington Heights, New York. From 1915 to 1962, he served as a dentist in the Lexington School for the Deaf, from which he had graduated, and the New York School for the Deaf in White Plains. In the rhetoric of the time, the Deaf community commented on Nies's attitude about communication: "Although an oralist bred, Mr. Nies takes to signs as a duck does to water, his delivery being as clear and forceful as if he were a manualist born: he has the advantages of seeing both sides of the shield and of being able to judge fairly, which cannot be said of the pure oralist or the rabid manualist" (*SW*, *24*[8], 1912, p. 139). Nies was also an instructor in oral hygiene for three years at Columbia University during the early part of his career, and for more than a decade he was a member of the staff of the Knickerbocker Hospital. His success illustrates the benefits realized as new professions began to open up during the last decades of the nineteenth century and the early decades of the twentieth century. Yet, while such opportunities for deaf people began to emerge in many occupations, the attitudes prevalent in society continued to be problematic, particularly among employers. The growing cadre of deaf professionals in the workforce, however, was developing an increasingly strong political voice; and deaf leaders appeared to be on a collision course with those who continued to treat them paternalistically.

3

Attitudes and Activism at the Turn of the Century

The growing involvement of deaf men and women in fields of science, invention, engineering, and medicine was overshadowed by the attitudes they faced in the society of hearing people. Through the years, a constant stream of advertisements for quack cures for deafness forced the Deaf community to always be on guard. Despite the comfort and confidence many deaf persons felt among themselves, outside the Deaf community they were treated as deficient. The pathological view of deafness dominated the thinking of the majority of hearing people and found expression in "cures" offered deaf people which included everything from inhaling chemicals to electrical stimulation, and which too many times were only vague promises.[1] This was well illustrated by the following advertisement placed in *The Silent Worker* in 1893 by Dr. George A. Leech of New York City:

> TO THE DEAF-MUTE WORLD
> It is gratifying to me, to be able to offer hope, to the multitude of the silent world, who are partially or even totally deaf. The day has arrived when this great affliction can be removed. By a recent discovery in connection with the Edison Phonograph, a method has been brought to light, through which hearing can be restored after a few months treatment. There are thousands who will not believe this, but there are ten thousand who will. This is offered to the majority. (*SW*, *5*[11], p. 3)

The pathological view of deafness was only exacerbated by Darwin's great synthesis, "descent by modification," which led to overenthusiastic and misguided attempts to establish legislation for breeding out "defectives" through genetic planning and social restrictions. Eugenics, the emerging science of applying knowledge about heredity to improve races and breeds, became an issue threatening the quality of life for deaf people. Although a long time brewing, the politicization of the Deaf community rapidly developed in the last decades of the nineteenth century, largely in response to the dilemma resulting from paternalistic attitudes of hearing people in influential positions. The eugenics movement, as it was applied to deafness, was capably led by a master of persuasion, Alexander Graham Bell. Like many others, Bell was captivated by Darwin's theory of evolution, which had led humankind to reflect on the theory's social and religious, as well as scientific, implications. Unfortunately, Bell's application of Darwin's theory caused much suffering among deaf people and their hearing relatives. The use of sign language and intermarriage of deaf people, in particular, became Bell's targets.

A related issue was the consequent diminishing participation of deaf people as teachers. Deaf men and women were dismissed from their positions in growing numbers because they were unable to teach speech to deaf children, and because their presence as teachers might promote further socialization among deaf people in the schools. This had an incalculable impact on vocational and technical education of young deaf people.[2]

Up until the turn of the century, the deaf fossil hunters and star watchers of note were practically consumed by their scientific interests and did not partake in the politicization of the Deaf community as it continued to grow. The National Association of the Deaf (NAD), only two decades old, had struggled against discrimination and prejudice; and the deaf scientists who served as leaders in these efforts were most often graduates of the schools for deaf children and Gallaudet College. These institutions, the foundation of the Deaf community, produced the chemist George T. Dougherty, one of the NAD founders, the microscopist James H. Logan, advocate for improved education and a superintendent of one school, and the botanist/entomologist Gerald M. McCarthy, who argued for better teacher training opportunities for deaf men and women in Gallaudet's Normal School. One of the greatest tragedies exemplified in the controversies of this era was the failure to systematically study the purported successes attributed by earlier educators to certain modes of communication with deaf people. This led to a continued beating of drums touting superior philosophies and instructional methods. This conflict and lack of systematic inquiry continues today.[3]

THE EUGENICS DEBACLE

When Anne Sullivan first came to Tuscumbia, Alabama, to work with young Helen Keller, she did so with Alexander Graham Bell's assistance. Bell had been in contact with the Kellers and Michael Anagnos, successor to Samuel Gridley Howe at the Perkins Institution for the Blind. During his first meeting with Helen, Bell fingerspelled Darwin's name into the deaf-blind girl's hand. When Helen asked him what Darwin had done, Bell enthusiastically responded that he had "wrought the miracle of the nineteenth century," and proceeded to explain to her how the *Origin of Species* had "widened the horizon of human vision and understanding" (Lash, 1980, p. 171). As one of the foremost proponents of applying eugenics to the eradication of deafness, Bell conducted numerous studies, published reports, and presented speeches. In his *Memoir upon the Formation of a Deaf Variety of the Human Race*, which he presented to the National Academy of Sciences on November 13, 1883, Bell discouraged deaf people from marrying one another, believing that this would reduce hereditary deafness. He concluded that congregation of deaf persons and the use of signs promoted intermarriage, and he urged that schools for deaf pupils be kept small and that instruction in articulation and speechreading should replace signs. Edward Allen Fay, the editor of the *American Annals of the Deaf*, was one of many who responded to Alexander Graham Bell's eugenic efforts. In Fay's *Marriages of the Deaf in America* (1898), he analyzed 4471 marriages and found that only about 9 percent of the children born to deaf parents were deaf. Fay concluded that the offspring of two deaf parents were no more likely to be deaf than those of one deaf and one hearing parent. The analyses of the marriages of deaf persons conducted by Bell and Fay came interestingly close to the Mendelian concept of dominant and recessive traits. But Bell's interest in eugenics reflected a greater concern for society than for individual human needs. To the Deaf community, Bell's *Memoir* was deeply offensive and threatening. As Winefield (1987) explained:

> Bell failed to comprehend that, to many deaf people, deafness was not an affliction, but a simple fact of life. Bell the scientist saw deaf people as different, as less well-equipped than those who could hear. It never occurred to him that there were some deaf people who were satisfied with their condition, who considered themselves normal, and who saw nothing wrong with having deaf children. To be told by a hearing person that they were "afflicted," and that none of them would want to pass on this affliction, was surely an insult. One may argue whether deafness is an affliction, but has anyone the right to make that decision for others? (p. 140)

Yet, in a paper titled "A Few Thoughts Concerning Eugenics," presented to the American Breeders' Association in 1908 and published in *The National Geographic Magazine*, Bell provided, along with an analysis of the principles of breeding, a warning of the inappropriateness of legislative restrictions:

> Can we formulate practical plans that might lead to the breeding of better men and better women? This is the great question we are called upon to consider. The problem is one of great difficulty and perplexity, for its solution depends upon the possibility of controlling the production of offspring from human beings. By no process of compulsion can this be done. The controlling power, if it is possible to evoke it in the interests of the race, resides exclusively with the individuals most immediately concerned. This fact, I think, should be recognized as fundamental, so that our process should be persuasive rather than mandatory. (p. 119)

In Bell's 1908 *National Geographic* report, he discussed the inalienable rights recognized by the Declaration of Independence to "life, liberty, and the pursuit of happiness," and stated that marriage often promotes such happiness, "even in the cases where the offspring may not be desirable." Bell wrote that the "institution of marriage not only provides for the production of offspring, but for the production of morality in the community at large . . . [this] is a powerful reason why we should not interfere with it any more than can possibly be helped" (p. 122). In one open letter to J. L. Smith published in *The Volta Review*, Bell responded to a protest by the National Association of the Deaf, concluding that "in order that you may know my attitude towards the subject of your communication, I may say that I have always deprecated legislative interference with the marriages of the Deaf" (*VR*, *10*[2], 1908, pp. 227–228).

Bell was a man of scientific inclination whose view of the world was influenced by Darwin. Married to a woman deaf since childhood, it did occur to him that individual happiness was an ultimate human goal. His commitments to inquiry and to an abstract kind of common good, however, were seemingly in conflict with the knowledge derived from his personal relationships.

The eugenics movement even affected the lives of deaf professionals who were neither "signers" nor involved in the Deaf community. On a "Record of Family Traits" developed by the Eugenics Record Office (Carnegie Institution of Washington) in 1925, for example, the astronomer Robert Grant Aitken was expected to describe his deafness.[4] The astronomer and physicist Annie Jump Cannon faced many attitudinal challenges to professional recognition, and there is evidence that her deafness may have been part of the reason that at least one important colleague rejected her for nomination for membership to the National Academy of Sciences (NAS). Raymond Pearl was one of the council members in the National Academy of Sciences responsible for considering whether the Academy should nominate its first woman member. Pearl was a biostatistician at the Johns Hopkins School of Hygiene and Public Health, a staunch eugenicist who was dissatisfied with the quality of recent candidates for membership in the NAS. Along with Edward L. Thorndike, Henry F. Osborn, John C. Merriam, F. A. Woods, and others, Pearl led the eugenics movement and was an active member of the Galton Society of New York. As the momentum increased among both the radical and conservative during the first few decades of the twentieth century, Pearl became a strong supporter of the use of birth control for eugenic purposes. As time went on, however, many scientists began to question the use and misuse of genetics. In 1927, Pearl himself appeared "anti-mainline," attacking the "biology of superiority" and claiming that eugenics had "largely become a mingled mess of ill-grounded and uncritical sociology, economics, anthropology, and politics, full of emotional appeals to class and race prejudices, solemnly put forth as science, and unfortunately accepted as such by the general public" (Pearl, 1927, p. 260). Yet only a few years before this publication we find in the private correspondence between Pearl and Edmund Beecher Wilson, a zoologist at the Harvard School of Public Health who was also interested in the problems of heredity, a tone which makes it difficult to evaluate Pearl's own prejudice. A few extracts follow and they well illustrate Annie Jump Cannon's plight as a deaf woman deserving recognition by a scientific society:

> March 7, 1923 Pearl to Wilson
> I note with some delight that two such reactionary birds as you and Henderson cannot bring yourselves to vote for a woman. Miss Sabin is certainly as good an investigator, if not better, than anyone of the other three names proposed in the same section, and in my opinion better than at least half the other nominees offered to the Academy this year.[5]
>
> March 9, 1923 Wilson to Pearl
> You are unkind. I know nothing about Miss Sabin. I thought she probably would have no chance to be elected this year as it was the first year she was up. . . . What you want to do if you want to elect Miss Sabin is to get the

> astronomers to run Miss Annie Cannon. I see no reason to suppose that she has not made real contributions to astronomy by the intelligence and intuition with which she has classified the Harvard spectral photographs.
>
> March 12, 1923 Pearl to Wilson
> I thought I would smoke you out by my remarks about Miss Sabin. I trust that you won't take my animadversions too seriously. My only objection to Miss Cannon as a member of the Academy is that she is deaf. It seems to me that in the passage of time as the Academy gets larger and near its limits, we might well introduce some statement about the undesirability of electing physical defectives. The astronomers seem to me to be a particularly bad lot. They are either deaf, blind, or spit, or something. I suppose the reason is that they are such great men that anything like an organic balance in their make up is not to be hoped for.

Whether or not there was an element of seriousness in Pearl's comment in his March 12 letter, we will never know. The NAS had elected several men who were deaf, and could hardly be accused of being restrictive in this respect. In fact, two of the Academy's incorporators, Frederick A. P. Barnard and his brother John, an engineer, experienced deafness, and Frederick was the Academy's correspondence secretary for years. Leo Lesquereux, profoundly deaf, was the Academy's first elected member. But Lesquereux, Frederick Barnard, and the deaf geologist Fielding Bradford Meek, elected to the NAS in 1869, had died many years earlier. At the time of Pearl's letter, there was at least one member in the astronomy section with a hearing loss, Robert Grant Aitken, and possibly others with physical or sensory disabilities. Although the attitude of Pearl about Cannon's deafness was not representative of NAS members, it is important to note that Cannon was never elected to the Academy. The nomination was the responsibility of the astronomy section and was never made. If Cannon or the anthropologist Ruth Benedict had been nominated, it is likely that such attitudes about their deafness would have compounded the already strong sexism and resistance toward women members during this era. Pearl's comments in these letters merely illustrate that this cannot be ignored as a possible explanation of why Cannon was not elected.

The Deaf communities in various nations continued to feel the pressure from eugenicists for years. In July, 1933, for example, as National Socialism was rising, the German government passed a new law for the sterilization of the "unfit," legislation which included hereditary deafness. Ernst Schorsch (1934) published a description of the law in *The Volta Review*. "We must not allow this burdening of our people with handicapped stock to continue," it was reported, and advisory offices were installed in schools for the deaf where card catalogs of all "deaf-mute" cases would be kept in order to determine those which were hereditary. "This would render a good service to eugenics" (p. 375). The report concludes with the statement that "naturally sterilization is an attack on the personal freedom of the deafmute [sic]. It must be remembered, however, that he is given special education training which makes it possible for him to enjoy his own home and the benefits of civilization" (p. 376).

As described in the *London Times* on March 21, 1935, as the controversy over sterilization continued to rage, Dr. Halliday Sutherland stood before the audience arranged by the National Council for Mental Hygiene in London and argued that sterilization was illogical, unscientific, antisocial, and of "no general utility at all."

While the eugenics movement lost its momentum over the years, today the issue of genetic research and the possibility of identifying deafness prior to birth raise many new concerns and controversies. In *Backdoor to Eugenics*, Troy Duster (1990) has warned against the growing trend toward molecular genetics technologies as possible answers to problems of specific groups. "The elimination or prevention of the 'defective fetus' is the most likely consequence and ultimate meaning of a genetic screen," he explained in his provocative analysis of the social, cultural, and moral issues associated with the revolution in molecular biology. "In a heterogeneous mix, the public forum for this debate needs to be vigorous and informed . . . the hour is late, the technology is closer, and the public debate has not been vigorous" (pp. 128–129).

RACISM AND PREJUDICE

Even within the Deaf community attitudes and prejudices presented some deaf people with insurmountable barriers. Racism was one such barrier. Most of the "Negro" or "Colored" schools in the South combined deaf and blind children together and, as late as 1950, thirteen states still had segregated programs for African-American deaf children (eight were still segregated in 1963). Julia Maestas y Moores and Donald Moores (1984) have indicated that special education in the United States has had a history of exclusion and discrimination with respect to Hispanics as well, and misdiagnosis and inappropriate labelings of Hispanic deaf children, in particular, along with a lack of Hispanic role models as teachers, continue to present a problem which borders on racism.

In *Blacks, Science, and American Education* (1989), Pearson and Bechtel described the societal barriers that have inhibited African-American participation in science. African-Americans have proven themselves innovative in the history of science. Lewis Latimer's improvement of the carbon filament, Norbert Rillieux's development of the process for making granulated sugar, Percy Julian's discovery of cortisone, and Edward Alexander Bouchet's work in geometrical optics are a few of the contributions of African-American scientists mentioned in this book. But, as in the case of deaf Americans in this era, barriers forced African-Americans to pursue industrial trades. As in much the same situation as deaf Americans have experienced, a shortage of African-American role models, inadequate recognition of achievement, and a lack of involvement of parents and teachers in motivating children are major themes in Pearson and Bechtel's analysis.

In 1950, Gallaudet College, eighty-six years after its founding, admitted its first African-American deaf student. In researching for their book *Black and Deaf in America* (1983), Ernest Hairston and Linwood Smith located a letter which not only reveals the problem of prejudice faced by African-American deaf people in the past, but also relates incidentally to the subject of this present work. Hume LePrince Battiste was a pupil at the Mt. Airy School for the Deaf who was interested in studying pharmacy. Apprehensive about attending Gallaudet College because of possible prejudice (he was probably Creole or Native American), he approached A. C. Manning, the school's superintendent, who penned the following inquiry to Edward Miner Gallaudet on August 24, 1908:

> My dear Dr. Gallaudet,
>
> I want to speak to you in behalf of Mr. Hume Battiste, one of our recent graduates at the Mt. Airy School. Last spring he took the College examinations and was admitted into your Introductory [pre-college] Class. But he is considered a Negro and there's the rub. The report is being circulated that the College boys intend to make it so disagreeable for him in case he enters college that he can't stay. He appeals to me for advice and I know of no better course than to lay the case before you.
>
> You know I am a Southern man with all the sentiments so frequently called "race prejudice", but I want to assure you that this young man is one of the most interesting boys I ever taught and so far is he above the average you naturally never think of his racial misfortunes. During the first year I taught him and found him to be an excellent character, possessing rare qualities—nor is he at all offensive in appearance or manner.
>
> His ambition is to become a pharmacist and I also desire to have you advise him on this point. Not being acquainted with a deaf pharmacist, I hardly knew whether or not to encourage him in this hope. His mother does not want him to attend College unless there is a possibility of his becoming a pharmacist as he already has a good trade at which he can make a living.
>
> I hope the young man may receive favorable consideration, for I believe he will prove himself worthy. (pp. 12–15)

Evidently, Battiste persevered, graduating from Gallaudet College in 1913. He held a position as a chemist for Carbon Company in Ohio, then studied briefly at the Oregon Agricultural College. At the start of World War I, he returned to Ohio to join the large workforce of deaf people employed by Firestone and Goodyear. He did not remain in the field of science after the war.

Today, deaf nonwhite scientists are nearly nonexistent. Christiansen (1987) has asserted that an estimate of 38,000 nonwhite prevocationally deaf people developed as a result of the 1972 National Census of the Deaf Population is likely an underestimation, and that no national data have been collected since 1972. He reported that in the 1950s less than 2 percent of nonwhite deaf males and virtually no nonwhite deaf females (90 percent of these persons African-American) attended college (as compared to 10 percent of white deaf persons). The segregated educational systems have contributed to this problem. In searching for deaf scientists of African-American heritage, I wrote letters to friends, colleagues, and authors of books about African-American scientists, and I contacted the Black Deaf Association (BDA). In the latter case, the BDA president, Carl Moore, knew of only one African-American deaf chemist, Alondra Little, at the Food and Drug Administration Pesticide Laboratory in Los Angeles. In addition, I located one deaf African-American mathematician, David James, a professor at Howard University. The inescapable conclusion from the research for this book is that deaf African-American scientists (the same can be said of Native American and other nonwhite deaf scientists) encounter so many barriers that they have yet to form even a small identifiable cadre.

THE WAR OF METHODS

The eugenics movement added fuel to the fire generated by the historical "oral-manual" controversy between persons advocating the use of speech and those supporting sign language in the education of deaf students. The beginning of this "War of Methods" is recognized as the dispute in the eighteenth century between Charles Michel Abbé de l'Epée, the "Apostle of the Deaf," and Dr. Samuel Heinicke, the "Father of German Oralism." Throughout history, bold and emotionally laden judgments regarding methods of communicating with deaf pupils have done little to bring the opposing camps together. There are many unpleasant instances in this tradition. One, in particular, was the International Congress on the Deaf held in Milan, Italy, in 1880, which was briefly discussed earlier in this book. Participants at this conference, overwhelmingly hearing educators, voted to proclaim that the German "oral method" should be the official method to be utilized in schools of many nations: "The congress, considering the incontestable superiority of speech over signs, for restoring deaf-mutes [sic] to social life and for giving them greater facility in language, declares that the method of articulation should have preference over that of signs in the instruction and education of the deaf and dumb [sic]" (Lane, 1984, p. 394). Many of the proponents of sign language communication were unable to attend, and deaf people themselves were excluded from the vote. The Deaf communities were understandably infuriated by the oppressive strategies of the hearing authorities in the schools, and the National Association of the Deaf (NAD) was established in the United States shortly after the conference in Milan in order to strengthen the political clout of deaf persons, who wished to have control over their own destiny. It was a human rights issue, in reality, and one that remains volatile.

This issue of communication was one that Edward Miner Gallaudet and Alexander Graham Bell argued over many times. Gallaudet was the champion of deaf people as they fought to have the "combined system" of spoken and sign language communication in instruction continued in the schools, and to preserve sign language. Bell broke away from the Convention of American Instructors of the Deaf to form his own organization, later renamed the Alexander Graham Bell Association for the Deaf, still active today, which advocated the teaching of speech and argued against the use of sign language. Bell, however, was a capable signer himself. In his "Fallacies Concerning the Deaf," he recognized the value of sign language: "To my mind it was the most interesting and instructive spectacle that has ever been presented to the mind of man—the gradual evolution of an organized language from simple pantomime" (1884, p. 51). Bell explained that in an earlier paper read before the Anthropological Society of London he had advocated the study of the "gesture language" by scientists, "for it seemed to me that the study of the mode in which the sign language has arisen from pantomime might throw a flood of light upon the origin and mode of growth of all languages" (p. 52). But Bell held to his own fallacy that "it is not the language of the millions of people among whom [the deaf person's] lot in life is cast. It is to them a foreign tongue, and the more he becomes habituated to its use the more he becomes a stranger in his own country" (Bell, 1884, p. 52). He described the "harmful" results of using sign language, that is, according to his belief, the difficulty in communicating at home or with friends who do not use signs and the weakened ties of blood and relationships that are a result of this. He felt that the residential institutions for deaf students would become their homes and that learning to think in signs would lead to difficulties in reading an ordinary book or in writing comfortably in English. Bell claimed that the extent of the deaf person's knowledge of the English language was the main determining cause of the congregation or separation of deaf people in adult life. But not far away at the Smithsonian Castle, Fielding Bradford Meek, a master of written English, was publishing many geological reports. A few years earlier, Leo Lesquereux, capable of reading and writing in French, German, and English, and an excellent speechreader, also had felt separated from the hearing community and did not bother to attend meetings of professional societies. The attitudes of the dominant society about deafness, not the language abilities of these deaf men, isolated them.

The involvement of deaf scientists in the War of Methods became much more visible when authorities of the New York Institution at Fanwood (now the New York School for the Deaf at White Plains) moved toward the use of the "pure oralism" approach. I do not mean to suggest that deaf scientists developed a definitive activism as a group. They have never been homogeneous in the political or even communicative sense, but at this time a number of deaf scientists joined ranks with other educated deaf persons in decrying such a change. Since they were successful as scientists, this added to their credibility and to their qualifications to criticize the educational establishment. Isaac Goldberg, a deaf chemist, was one of the most outspoken. He was born July 1, 1865, and graduated from Gallaudet College in 1888. For fifteen years he progressed in several businesses, including the position of chief of the Chemical Branch of Frederick Loeser and Company in Brooklyn. There he was in charge of the analysis necessary for manufacturing drugs, medicines, flavoring extracts, and other goods, and of assisting the company in lawsuits related to the quality of these products. Upon receiving the news of the school's plans, Goldberg wrote to Enoch Henry Currier, the principal of the New York Institution on January 31, 1912: "I have just learned with surprise and sorrow, not for myself, but for the coming generation of the deaf, that the authorities are considering the question of installing Pure Oralism in your Institution and that there exists a possibility of the change being [consummated]. Evidently the authorities intend to disregard the deaf themselves in this matter" (Currier, 1912, p. 5). Goldberg described his own experiences in being educated in the "foremost Oralistic School of the day," and how lectures and chapel services "were never held while he was a student." As he explained, since signs were not permitted and since they were "the only proper medium for delivering a lecture to the deaf," "the matter was dropped entirely, to the great detriment of the pupils" (p. 7). In addressing this same issue of the difficulty in communicating through speechreading only in public presentations, Amos G. Draper, a deaf professor of mathematics at Gallaudet, referred to the articles he had published in the *American Annals of the Deaf*: "I can write no better now than then" (p. 18).

With courageous resistance many deaf people wrote from across the country. F. R. Gray, the deaf lens maker and astronomer from Pittsburgh, wrote in support of the combined method: "I have had personal opportunities of judging at first hand of the effects of the pure oral method, and from all I have learned, hardly any greater injury could be done the deaf as a whole than to abolish signs from the school room. . . . Be sure that though you may not hear directly from them, the deaf in the United States are solidly with you, and if called on will back you up in any way they can" (Currier, 1912, pp. 9–11). George Dougherty was then chief chemist of American Steel Foundries in Chicago. Writing on company stationery, he too took a strong stance in favor of the combined method: "Your board of directors will be doing the wise thing if they would give more weight to the uniform, universal verdict of the adult educated deaf, who have been 'through the mill' of school and world experience" (Currier, 1912, p. 37). The inventor Anson R. Spear described his success in the business world, where "the real test of the efficiency of a school for the deaf—of methods of education—is to be found" (p. 15). Spear strongly believed that the "combined system, where signs are used, produces the best results" (p. 16). Another deaf inventor, Anton Schroeder, asked, "What right do they have to take our happiness and privilege away from us?" (Currier, 1912, p. 80). Francis P. Gibson, a third inventor and the secretary of the National Fraternal Society of the Deaf, an insurance organization, wrote, "Why, your Board might as well take away books" (Currier, 1912, p. 25), and the deaf chemist Ernest Dusuzeau wrote from Paris: "It is good to teach speech to a deaf-mute [sic], but to complete his education, speech is not so great an aid as the gesture language, the only language which will brighten his comprehension. How can you teach him literature and the sciences? The best method of instruction is unquestionably the *mixed method*, which is to say the oral method and the sign method combined together. . . . I will not conceal that it is to the sign method I owe my education and my position, and that it is to it that I am indebted to receiving Bachelor of Science at the Sorbonne" (Currier, 1912, p. 75).

These were typical of the many letters mailed to Currier. I have quoted several scientists, but letters were also sent by deaf men and women artists, poets, and leaders in many other professions. In 1912, Currier published the collective views of these deaf professionals in *The Deaf: By Their Fruits Ye Shall Know Them*. The convincing arguments he received were summarized in the preface of his book and reveal his insightful conclusion: "A perusal of the letters presented will confirm the unprejudiced reader in the belief that all repression must be eliminated, and that the educator of the deaf must learn through the experience of the educated deaf wherein to modify and improve his methods" (p. iv). He received public appreciation from the Deaf community for his efforts. "Professor Currier has rendered the cause of education a notable service," stated J. H. Cloud, "and won the lasting gratitude of the deaf of the world by publishing in pamphlet form numerous letters of protests against the proposed invasion of 'pure oralism' in his school" (*SW*, *25*[6], 1913, p. 101).

Many years before Currier's book was published, Luzerne Rae had eloquently presented his valuable perspective that there is no greater obstacle to progress in art or science than a bigoted and unreasoning attachment to a system. Applying this to communication with deaf people, Rae believed eclecticism was the only sound philosophy: "The only proper position for a true man to take is one of perfect independence; whence he can look, with an equal eye, upon whatever comes before him, and receive every applicant for his favor, precisely according to the proofs of positive value which it brings" (*AAD*, *5*[1], 1852, p. 22).

4

The Twentieth Century: The First Fifty Years

TECHNICAL AND VOCATIONAL EDUCATION: AN UNFULFILLED NEED

When the National Association of the Deaf (NAD) was organized for purposes of self-advocacy in 1880, on the heels of the Industrial Revolution, "industrial education" was a priority issue to the Deaf community. It was the opening topic on the agenda for the first NAD convention on August 25 of that year. Higher education in technical areas for deaf people in the United States was also an issue of importance in the Deaf community before the turn of the century. At the Conference of Principals and Superintendents of the American Schools for the Deaf held in Colorado Springs in August, 1892, a proposal for the establishment of a "national technical training school for the deaf" was presented for discussion by F. D. Clarke from the Arkansas School for the Deaf. In September, *The Silent Worker* published a brief but provocative editorial comment: "By all means let us have a technical school. It is just as important as [Gallaudet] College at Washington. We should say that it is more important" (*SW*, *5*[6], 1892, p. 4). Proposals proliferated, including a suggestion by Warren Robinson, a deaf leader, that a "technical department" be added to Gallaudet College:

> We mean that the time has evidently arrived when technical training ought to be given those who desire to follow other callings than that of teaching. To this end such departments as those of Civil and Mechanical Engineering, Electricity, Business, etc., should be created, thus giving to the College something of the character of a university, by means of which students may prepare themselves to step from the College directly into some profitable employment, instead of having to go elsewhere for preparation. (*AAD*, *37*[1], 1892, p. 30)

Robinson's article calling for a technical school aroused interest. In the following year, the Conference of Principals and Superintendents of the American Schools for the Deaf passed a resolution to "do what they could" about establishing a technical department. In 1895 Gallaudet College reported that the technical department, "which is soon to become part of the college," with opportunities to study architecture, practical chemistry, electrical and mechanical engineering, surveying, and related areas, "is exciting considerable comment" (*B&B*, *3*[4], p. 52). But comment did not translate at a satisfactory rate to fruitful programs and, in subsequent years, calls for improved technical education for deaf people in the United States and other countries continued (Waring, 1896). Efforts included W. W. Turner's paper "High School for the Deaf and Dumb" and the NAD's repeated pleas for further attention to this need, both in existing school programs and in higher education environments. In 1900, at the Chicago Chapter of the Gallaudet Alumni Association Annual Dinner, the deaf metallurgist George T. Dougherty belittled the technical department as it was then conducted at Gallaudet College, declaring that "to be effective, it must have a special corps of instructors" (*B&B*, *8*[7], p. 311). In 1904, Isaac Allison once again raised the issue of technical education in the *47th Annual Report of the Columbia Institution.* Years went by with still no satisfactory progress, prompting several deaf leaders to express concern in 1912. The deaf inventor William E. Shaw wrote a letter to the editor of *The Silent Worker*: "There is a very great need of a Technical Institution for the deaf somewhere and I sincerely hope the deaf especially will be interested enough in electricity to help in any way to further such a movement for an institution of this kind" (*24*[8], 1912, p. 151). And Warren Robinson continued to encourage others to work toward this end: "The question now is, are the deaf keeping abreast of the changes that are going on in the industrial world and taking advantage of all the opportunities offered them? Hardly" (*SW*, *24*[8], p. 135).

Meanwhile, some deaf men and women interested in a technical education for themselves were struggling without support in regular university courses. We find most of the reports of their successes, however, not focused on the opportunities for technical education but rather on the rhetoric of the communication controversy. While scientists and science students with "oralist" preferences did not partake in the dispute over methods which occurred in 1912 at the New York Institution, one need only read the early issues of *The Association Review* to find their own efforts to advocate a communication method. This journal, renamed *The Volta Review* in 1910, was published by the American Association to Promote the Teaching of Speech to the Deaf (now the Alexander Graham Bell Association for the Deaf). The first issue of Volume 1 contained an autobiographical sketch of Alexander L. Fechheimer, who was deafened while an infant. Fechheimer claimed that his speechreading abilities were responsible for his success in a program in architecture at Columbia University which included calculus, mechanics, and other science courses. Following his example, Hypatia Boyd, describing her university experience as a "stranger in a strange place" (*VR*, *2*[2], 1900, p. 130), provided an autobiographical sketch which included a description of her endeavors in science. Boyd was deafened at the age of six and a half and is one of the earliest profoundly deaf women to have enrolled in science courses in a university for hearing students. In emphasizing her "oral" education as a factor, she presented the following anecdote pertaining to speechreading in her essay titled "University Experiences":

> And this reminds me of a noble and unselfish act by one of my teachers. He was anxious to have me read his lips, but finding that I made very little progress, inquired the why and wherefore of such backwardness on my part. As might be expected, I disliked telling him the real cause, and yet I could not state anything but the terrible truth. And so, summoning up all my courage, I explained to him that his moustache was of the grievous, heavy, overhanging sort, so as to completely conceal his lips, thus rendering lip-reading an impossibility in his case. Imagine my feelings then, when a few days later, he appeared in class with a goodly portion of his moustache removed. Thereafter, I no longer felt tempted to use my scissors lavishly, as I got along in Science very well, you may be sure. (p. 129)

Interestingly, Boyd married a man who did not speak or speechread and who was a fluent signer. She learned to sign herself, took a position at the Wisconsin School for the Deaf, and published regular columns in *The Silent Worker* for the Deaf community.

Another example of this rhetoric is found in the report about the four deaf men who graduated from Harvard's Lawrence Scientific School in 1902. Homer Wheeler, his brother Melvin, and Robert Pollak, were congenitally deaf and majored in civil engineering. Tilston Chickering, described as "partly deaf," completed his course in mechanical engineering. In *The Evening Trumpet*, published in Boston, the skills of these students in speechreading were highlighted, while no mention was made of their academic abilities or fortitude:

> For the first time in its history, it is believed, Harvard College has bestowed a regular degree on a deaf-mute [sic]. This week four young men afflicted in this way, two of them brothers, were graduated. The four are all entered in the Lawrence Scientific School, and all are planning to be engineers. . . . Their eyes have done double duty, the slightest motion of their instructors' and fellow students' lips being full of meaning to them. (*VR*, *7*[1], 1905, p. 28)

Professor James Love of the mathematics department was given special credit because "in order that they might better read his lips, [he] cheerfully sacrificed his beard" (p. 29).

It was a rare occasion when even an "orally-taught" deaf person could instruct students in articulation, and for this reason the scientifically inclined deaf individual, regardless of preference of communication strategies, was basically precluded from entering teaching during the early decades of the twentieth century. Unable to teach articulation, many talented deaf teachers had already been disqualified from positions in schools which took up the banner of the Edict of Milan. Alexander Graham Bell had added to the fire of discontent when he further argued that nearly one third of the teachers of deaf children in America were themselves deaf and that "this must be considered as another element favorable to the formation of a deaf race—to be therefore avoided" (1883, p. 48).

The first deaf science teacher at Gallaudet College was Julius J. Heimark who joined the faculty in 1913 and taught both chemistry and biology; and the first deaf mathematics teacher at Gallaudet, Frederick H. Hughes, was hired in 1916. Neither had graduated from Gallaudet's Normal School. In a letter written to President Edward Miner Gallaudet on May 27, 1901, the deaf entomologist and botanist Gerald M. McCarthy criticized the policy of not awarding fellowships to deaf persons in the "Normal Department," which educated future teachers. He feared the consequences of such exclusion. McCarthy's letter reveals common threads which we find interwoven in the lives of deaf scientists who graduated from Gallaudet College in this era. First, there is a deeply felt appreciation for and an attachment to the college, which had graduated them. Second, the Deaf community monitored the progress of the college as it promoted a better quality of life and work for deaf persons in general. And third, for many graduates, the college represented the Deaf community's link to higher education, its gateway to societal change, and the potential for making deafness and the contributions of deaf people visible, particularly with regard to leaving a distinctive mark on other institutions for learning. "I cannot see," McCarthy wrote in his letter to President Gallaudet, "how a college specially and technically established for the deaf can legitimately use its funds to educate and by pensioned fellowships maintain, students who are not deaf—the graduates of the college itself being excluded from these fellowships. . . . I cannot but feel that the 'normal' expansion of our old college has been very abnormal, and contrary to the spirit of its founders."[1] Gallaudet College did not open its programs for the master's degree in education to its own graduates until 1963.

The loss caused by the Edict of Milan, complicated by the attitudes of the eugenicists, was a tragic one. The number of deaf teachers in the school programs eventually dropped to about one in ten by the 1920s and has not significantly changed since. Qualified deaf men and women who might have capably taught in technical education classes were left to find other means of employment.

THE FIRST WORLD WAR AND THE GROWTH OF INDUSTRY

When President Woodrow Wilson capably dispatched his plans to increase agricultural and industrial production to meet the demands placed on the United States by its entrance into World War I, the inventive talents of many hearing and deaf people were applied. Fuaad Jerwan, a graduate of the Clarke School for the Deaf in Massachusetts who had earlier begun taking out patents for "hydro-aëroplanes" at Fort George, New York, used his skills to assist Lieutenant Blair Thau, a pilot of the Lafayette Escadrille who was later killed in France. In 1917, William E. Shaw submitted his blueprints for a "robot bomber" to the U. S. Navy Department, but aeronautical experts considered him a "visionary." More successful was Richard E. Dimick, who began to envision other applications of Marconi's work with radio. The secretary of the navy was greatly interested in Dimick's idea of developing a device to locate submarines, and appropriated $30,000 for his research. The deaf inventor pursued the project day and night, although he was confined to a wheelchair by repeated attacks of appendicitis. The device Dimick designed was successful and, as Round (1929) reported, the navy put it into immediate operation just as the United States had begun to transport troops to France. Dimick had died, however, and he never saw how he may have contributed to saving the lives of those at sea.

By the time World War I began, chemistry had become the frontier scientific profession for deaf persons in terms of both occupational statistics and visibility in the Deaf community. This was the result of increasing demands for industrial chemists and the result of Gallaudet College's solid chemistry curriculum (much stronger than in other areas of science). As the chemist Oscar Guire described in his "Sketches of School Life" (*DA*, *19*[9], 1967, p. 30), Gallaudet in the early 1920s offered no substantive curriculum in physics: "During my first year at Gallaudet College I saw that chemistry was the only thing for me. All the college offered in physics was (1) one term (3 months) of mechanics which was required of all freshmen, (2) one year of general physics which was required of all sophomores, and (3) one term of electricity which was optional for juniors and seniors." Guire went on to describe the chemistry curriculum, explaining that physical chemistry was added in his senior year and his class was required to take it for three terms.

At this time, chemistry was also the leading field of science for hearing men. But there the similarity ends. While approximately 15 percent of hearing scientists were in chemistry, and about an equal number were in the medical sciences (Rossiter, 1982), a conservative estimate for the distribution of deaf scientists in chemistry in 1921 would be about 70 or 80 percent of all deaf people in scientific fields. The actual number of deaf chemists in 1921, however, was likely not more than thirty (compared to 1350 hearing chemists).[2]

The improvement of the Gallaudet curriculum since William Hill's description of his experiences in the 1860s was especially obvious, as indicated by a report published by the deaf industrial chemist Isaac Goldberg. Goldberg's employment experience included research for the Gloria Products Corporation for research on a method for treating the leaves of plants to give them the gloss desired for wreaths and other funeral decorations. His process for doing this was favored over those of other chemists, and the company offered him a partnership. During subsequent years, he conducted research on the catalytic action of aluminum chloride on spent crude oil in a process for the production of gasoline and kerosene oil. His enthusiasm for his work was shown in his publication in the Gallaudet College organ, *The Buff and Blue*, in which he encouraged deaf students to consider chemistry as a career:

> It is my firm belief, borne of an experience covering a period of over thirty years, that deafness in no way militates against the highest success in the field of chemistry and I am satisfied the opportunities for learning the science of chemistry at Gallaudet College is fully on a par with any other institution of learning, especially for the deaf, provided always the student puts the best that is in him to the tasks that the study requires. Gallaudet College afforded me all the chemical schooling I ever had and it appears to have been adequate to enable me to occupy a position that had hitherto been filled only by big university men; consequently by taking full advantage of one's time and opportunities at Gallaudet the diligent student can assure himself of all the training necessary to a successful chemical career. (*B&B*, *29*[2], 1920, p. 7)

World War I was an important factor in bringing business managers to realize the value of deaf laborers and professionals. As hearing men went to war, more industrial positions were offered to deaf people. In 1920, G. C. Farquhar summarized the influence of this trend on one field of science in his article "The Deaf in Industrial Rubber Chemistry": "In the growth of the rubber industry which led to the development of the phenomenal Silent colony in Akron, Science has been Labor's co-worker and in the laboratory the crude experiments of Charles Goodyear have been elaborated. . . . And the deaf are represented there" (p. 151). Kreigh B. Ayers was one of the first deaf chemists hired by Goodyear as the United States entered the war. He was deafened by spinal meningitis at the age of two and was educated at the Ohio School for the Deaf in Columbus. After leaving Gallaudet College in 1915 without earning a degree, Ayers was hired as a chemist for Westinghouse Company and the National Carbon Company in Cleveland, conducting analytical work and producing chemicals on a large scale for the carbon industry. When a hearing chemist from the organic department of Goodyear was sent to war in 1917, Ayers was asked to replace him. This was considered an honor, since the department conducted confidential work and "only men in whose loyalty, discretion, and skill the company places implicit trust are chosen for its work" (*B&B*, *26*[9], 1918, p. 343). Ayers' knowledge of electrochemical processes was in enough demand that Goodyear, for which he worked twenty-five years, loaned him to their Aircraft Corporation. There he published various reports on the solubility of sulfur in rubber, and he was considered to be the "best of the dozen deaf chemists in America" in the early 1920s (*SW*, *34*[6], 1922, p. 238). Ayers was also a leader in the local Deaf community. He organized the famous "Goodyear Mutes" football team that competed for popularity with the Akron pros, and he was the team's manager in 1918 and 1919, when it became one of the best-known gridiron organizations in the history of Akron.

Two Gallaudet graduates in chemistry who were quickly offered positions by the Goodyear Testing Laboratory when the United States entered the war were Foster D. Gilbert and Clifford M. Thompson. Gilbert graduated from Gallaudet in 1917, began in general testing at Goodyear, and was promoted to special testing of materials such as antimony and zinc. Thompson was educated in the Utah, Colorado, and Idaho schools for deaf children and graduated from Gallaudet College in 1916. At Goodyear he specialized in the study of "accelerators" used to hasten the process of vulcanization of rubber. At one point the laboratory supervisor was not satisfied with the quality of chemical analyses being conducted and gave the entire testing staff a sample with a percentage of sulfur. The only chemist whose report showed an accurate determination was Thompson.

The Goodyear Mutes football team was representative of the interest the company held in hiring deaf people. Noting the success experienced by Goodyear with deaf workers, Firestone also expanded its own force of eight deaf men, announcing its interest in a "new silent colony" headed by a deaf director, Benjamin M. Schowe (G. C. Braddock, 1919). By 1919, Goodyear and Firestone employed over five hundred deaf people. In many ways, these companies expressed support for their deaf employees; and, as the Deaf community around these companies grew, social, athletic, literary and other groups and activities developed, and the relationship the companies held with the deaf men and women in their employ was used by the Deaf community as an example for other businesses. Gotthelf (1919) described how the Firestone Company treated its deaf employees "as they do the hearing," evidenced by a banquet held at the Firestone Clubhouse, at which the deaf employees were guests. "This was given to show the company's appreciation of their services." Henry Ford was also supportive of deaf workers and, when unable to do well in one department, they were given other opportunities (*SW*, *33*[2], 1920, p. 69).

Many deaf workers were general laborers, but Firestone also hired Thomas W. Osborne, who spent his entire career as a research chemist there. Osborne began as a junior chemist and was eventually promoted to the status of senior analytical chemist and then assistant to the director of analytical laboratories. He lost his hearing when he was four years old from measles and diptheria and attended the Tennessee School for the Deaf in Knoxville, graduating in 1914. In 1919 he received a B.S. degree from Gallaudet College. After moving to Ohio, he studied colloid chemistry at Akron University and took courses in chemical microscopy and microchemistry at the Case School of Technology. Osborne organized the first microanalytical laboratory in the Firestone Research Laboratories, for which he was later honored.

With the arrival at Firestone of Claude V. Ozier from a perfume firm in Tennessee in 1920, the number of deaf chemists in that town rose to five and there was talk of the formation of a "chemists' club."

While industrial rubber research and World War I were catalysts for employing early deaf chemists, there were many other chemical processes rapidly gaining attention, including the production of synthetic fertilizers, explosives, dyes, polymers, and drugs, and in this context we find the remaining deaf chemists of this period employed in private businesses, city assay offices, hospitals, and early enterprises of their own. Daniel C. Picard, profoundly deafened at the age of nine, attended a public school in New Orleans for three years, then completed his education at the Louisiana School for the Deaf. During this period he had special tutoring in preparation for Gallaudet College, from which he graduated with a bachelor's degree in 1899 and a master's degree in 1900. Picard also earned a bachelor's degree from MIT. He was first hired by the Gate City Cotton-Seed Oil Mill in Egan, a suburb of Atlanta, Georgia, where he gained favor with his employers by discovering an error in other chemists' analyses of cottonseed meal by which the company had been losing thousands of dollars. In 1913, he established his own successful laboratory and later consolidated with another laboratory owned by Thomas C. Law. The new business, Picard-Law Company, specialized in the analysis of seed products and fertilizers, with offices branching into many other cities. Picard was honored for his pioneer work in cottonseed products.

The development of industrial research was the primary factor explaining the rise of chemistry as a profession for deaf people. But these early deaf chemists held far from ordinary careers in chemistry. David Friedman, for example, an immigrant from Hungary who was deafened by scarlet fever at the age of seven, graduated from the Ohio School for the Deaf at Columbus in 1900 and received a B.S. degree from Gallaudet College in 1904. Friedman worked in the chemical laboratory of the City Health Department in Cleveland for twelve years, conducting analyses of samples and testing paving materials, a rather routine experience. Noting his unobtrusiveness, officials recruited him to assist them in cleaning up a drug ring. To accomplish this, he responded to a newspaper advertisement and his subsequent work resulted in the arrest of several drug peddlers. Friedman then became involved in trapping doctors practicing without licenses. When the risks of such assignments became too high, he left the city position to pursue chemical work full-time again. He spent thirty years at J. L. and H. Stadler Rendering and Fertilizer Company in Cleveland, where he conducted analytical work in the rendition of fats and the manufacture of fertilizers and animal feed. James W. Howson, totally deafened at the age of nine, enrolled in the California School for the Deaf at Berkeley as a day pupil when his parents moved to Sacramento. He earned a bachelor's degree and master's degree from the University of California. His first position was with the Union Sugar Company in San Francisco. Howson was called to Nevada to testify at the trial of two miners caught with stolen gold ore. He verified the ore's source through a chemical analysis of its bismuth and tellurium content. At similar trials, convictions of indicted miners invariably followed Howson's testimony. Later, while a chemist in an assay office, he joined the faculty of the California School for the Deaf, where he taught physics, chemistry and other subjects until his death in 1941.

THE YEARS BETWEEN WORLD WARS

Of the various forces shaping the quality of life of Americans in the decades following World War I, radio, moving pictures, and the automobile played dominant roles. Without any access to the telephone, long-distance communication remained tedious for deaf people. The telephone had inestimable value to hearing people in emergencies, and, for social discourse and business transactions, it was a boon to humankind. Deaf people enjoyed no such luxury. One need only give up use of the telephone for a few days to appreciate the handicap the invention introduced into the lives of deaf people. The frustration of one young woman, Katherine Steffens, is evident in the report of how she had taught herself to "hear" the telephone by placing her fingers inside the earpiece to translate words she felt through the diaphragm. The validity of this report about the "stone deaf" Steffens is questionable. "The nerves and skin of the young woman," it was explained, "are extremely sensitive and she experiences no difficulty" (*SW*, *35*[2], 1922, p. 59). The Saskatchewan Government Department of Telephones, experimenting with the use of amplification of the intended signal combined with suppression of extraneous sounds, reported an equally sensational statement: "Some persons who have been unable to hear in 20 years have transacted 'long distance' business over the phone." In another report (*SW*, *36*[10], 1924, p. 482), "deaf" people were purportedly able to hear the telephone through their cheekbones.[3]

Radio was rapidly becoming a new medium for entertainment and education, but this technology further disadvantaged deaf people. The "silent movies," enjoyed by the Deaf community on an equal basis with hearing friends during these early years, were being replaced by the "talkies." Hence, another technological breakthough that brought educational benefit and pleasure to hearing people introduced a handicap to deaf persons. In the magazine *Motion Pictures* for May, 1929, Charlie Chaplin complained that the talkies would ruin "the great beauty of silence": "They are spoiling the oldest art in the world—the art of pantomine" (Batson & Bergman, 1985, p. 303). In the same year Edison also expressed concern about the talkies. He told newspaper reporters at his winter home in Fort Myers that inventors should be called upon to perfect an apparatus to enable deaf people to enjoy motion pictures again. "And it looks as though it is another job in store for me. Now that they are turning the movies into 'talkies,' it lets me out for I can't hear a thing" (*SW, 41*[5], 1929, p. 217). Various efforts were subsequently made to help deaf people continue to enjoy films. Edison's "Talkies for the Deaf" were a prelude to captioned films. In 1937, Ernest Marshall produced a motion picture in sign language for deaf audiences. Some theaters experimented with wires connected to microphones located near the loudspeakers behind the cinema screen, and in 1939, the use of a "telesonic system" was reported consisting of a pickup coil laid around Wyndham's Theatre in London which freed a person with deafness from being tied to the seat (*Nature*, April 15, 1939, p. 633). But these efforts were not satisfactory for deaf people, and the disadvantage they experienced remained an emotional issue. As one deaf person complained in *The Lancet* on March 12, 1938, all films should include captions. "The coming of talkies has been a serious blow to the deaf," the correspondent wrote, "many of whom have lost all pleasure in this form of entertainment and instruction" (p. 648).

Even the use of the automobile by deaf people was restricted, even though it did not require hearing to operate or enjoy, was restricted, and this became a legal issue to be dealt with by the Deaf community. In Maryland, for example, a law required a deaf driver to be accompanied in the front seat by a person with normal hearing.

During this period, the Deaf community saw the promise of rehabilitation assistance begin, first with the federally funded efforts in the form of the Smith-Fess Act in 1920 and then with private organizations such as the National Rehabilitation Association. In 1926, the *Rehab Review* began publication through the efforts of the state directors of the New York and New Jersey vocational rehabilitation programs. By the time the First World War had ended, the potential of people with disabilities on the production line as well as in other positions in government had been recognized, and many of these workers were kept on. Men returning from the military established groups such as the Disabled American Veterans; but the NAD, with its scarce resources, was unable to seize the opportunity to provide rehabilitation assistance to deafened veterans in a significant way.

Moreover, little had yet been done about improving technical education for deaf people and, in 1930, Peter N. Peterson, a deaf man from the Minnesota School for the Deaf repeated the plea made two decades earlier by the electrical inventor William E. Shaw. A "national technical institute for the deaf," Peterson wrote, "located in a large manufacturing city, is what deaf young America needs more than anything else" (Peterson, 1930).

Chemistry Continues to Lead

After the First World War, the mix of graduates of Gallaudet College and other deaf Americans who had found success in colleges and universities for hearing students continued to expand, and the list of deaf scientists in the United States grew, albeit not at an impressive rate. Those scientists who had graduated from regular universities had minimal academic support, usually from classmates who shared notes from lectures. There was also no assistance from sign language interpreters, who had yet to form a profession. In addition, there were men deafened during the war; only a few mentioned in the literature of the Deaf community. One was O. R. Sweeney, who had lost his hearing while working in the arsenal at Edgewood, Maryland, as a result of poison gas. Sweeney attended Ohio State University at Columbus, where he received a B.S. degree in chemical engineering. His Ph.D. was earned at the University of Pennsylvania, and he did postdoctoral work at Göttingen, Germany. Using an "ear phone" (Acousticon), Sweeney taught at Ohio State University, Pennsylvania University, Cinncinati University, and the Agricultural and Mechanical College at Fargo, North Dakota. He was head of the Chemical Engineering Department at what was then Iowa State College in Ames, Iowa. Sweeney was a pioneer in the water-softening industry and helped to perfect the zeolitic softening method. He also developed over two hundred products from corn stalks and cobs in his experiments on commercial utilization of agricultural wastes (Marnette, 1927). Another veteran was Walter Bell, deafened while serving with the Canadian Field Artillery in Europe. He eventually became director of the Geological Survey of Canada. Bell was somewhat withdrawn because of his deafness, and "not at his best in large gatherings" (H. N. Andrews, 1980, p. 261). With the objective of using fossil plants as a service agent in stratigraphic geology, he produced many publications about plants based on his field experience, particularly in the coalfields of Nova Scotia and New Brunswick.

As World War II approached, more deaf people continued to enter chemistry, while far fewer chose biology or physics. In the perspective of one biographer from this period of time, the milieu appeared promising. "Half a century ago the deaf chemist was a rarity, worth notice, in the Almanac of Strange Facts," wrote Braddock in 1938, "but nowadays there are so many of them that no one could furnish a complete list offhand, and these eminent individuals occasionally suffer from lack of public appreciation" (Braddock, 1975, p. 200). While it is true that educational and employment opportunities were improving for this generation of deaf scientists, a list of them would likely still not be very long. One may find in the brief reports about these scientists an increased involvement in both scientific and community organizations, and improved opportunities to take supervisory positions. Deaf scientists in the preceding decades were hired primarily into industrial positions and small businesses. Now, a trend toward employment by agencies of the federal government had become evident. Increasingly, Gallaudet graduates were hired into positions by the U. S. Department of Agriculture, Department of Commerce, Navy, Army, Office of Health, Education and Welfare, Food and Drug Administration, and Bureau of Standards. As in the earlier years, barriers faced by deaf scientists were often created out of ignorance. In one case, Wilson H. Grabill, who eventually became chief of the Family and Fertility Statistics Branch of the Population Division in the Bureau of Census and a coauthor of three books on fertility statistics, first had to challenge the civil service examination requirement which stipulated that applicants be able to hear a watch tick at a distance of fifteen feet (A. B. Crammatte, 1987).

Representative of the growing cadre of deaf scientists from Gallaudet was Oscar D. Guire, who lost his hearing through scarlet fever at the age of five and graduated with a B.S. degree from Gallaudet in 1921. While studying for a master of science degree in chemistry at the University of California, Guire worked for Professor Paul Hibbard of the Agricultural Chemistry Plant Nutrition Division, College of Agriculture, where he assisted him in analyzing samples of water and soil sent by farmers and land buyers, and in testing some analytical processes involving water and soil. He completed his master's degree in 1923 and was an analyst for the California Portland Cement Company for twenty-two years. In 1947, Guire was partially paralyzed through a cerebral hemorrhage. He joined the staff of the American Chemical Society (ACS) in 1949 and produced about three hundred chemical abstracts per year. In 1964 he was highlighted in an ACS publication for his years of service as a chemical abstractor.

Other deaf scientists graduated from universities and colleges for hearing students. Edward Croft, for example, born in the Philippines, was the son of Major General Croft, chief of staff in infantry during World War I. He attended the California School for the Deaf, Leavensworth High School in Kansas, and the Polytechnic Preparatory School in Brooklyn. He graduated with a bachelor's degree from Union College in 1929, received his master's degree in chemistry from the Massachusetts Institute of Technology in 1932, and then was employed as a chemist in the U. S. Chemical Warfare Section of the War Department in Washington, D. C. H. Latham Breunig lost some of his hearing at the age of three. Scarlet fever and an accident eventually left him completely deaf. He attended the Clarke School for the Deaf in Northampton, Massachusetts between 1920 and 1927 and Shortridge High School in Indianapolis. At Wabash College in Indiana he received a B.S. degree in chemistry in 1934. His Ph.D. was earned four years later from Johns Hopkins University. Breunig began full-time research work in chemistry at Eli Lilly and Company in Indianapolis after graduation. He was the first deaf president of the Alexander Graham Bell Association for the Deaf.

Biology and the Medical Sciences

Three deaf persons who contributed in very different ways to the field of biology are briefly mentioned here. Regina O. Hughes, a graduate of Gallaudet College in 1918, began as a translator for the U. S. Department of State, one of only two non-French women employed by the U. S. Office of Translation. Throughout her productive career, thousands of her scientific illustrations have appeared in textbooks and publications. Her botanical paintings and drawings are exhibited in many museums, including the Brookside Gardens (Maryland), the

Selby Botanical Garden (Florida), the Hunt Institute for Botanical Documentation at Carnegie-Mellon University in Pittsburgh, and at Gallaudet University. A collection of forty water colors of orchids was exhibited in the Rotunda Gallery of the Museum of Natural History of the Smithsonian Institution. Six thousand of her illustrations appear in one work, *Economically Important Foreign Weeds: Potential Problems in the United States* (1927), a U. S. Department of Agriculture handbook which took two and a half years to complete. For all of these illustrations she wrote the plant descriptions. Hughes' experience has included years of honored illustration work for the Agricultural Research Service and the U. S. Department of Agriculture. She has also been a resident illustrator in the Department of Botany at the Smithsonian Institution. Her illustration of a bromeliad was on display in an exhibit of the research of Lyman B. Smith and Robert W. Reed. In 1979, the new species illustrated in this painting was named *Billbergia regina* in her honor by the Smithsonian Institution. Interestingly, Hughes is the only person to have both a plant genus and species named for her.

Hubert Lyman Clark (1870–1947) was thirteen years old when he published his first two papers on butterflies. During a field trip to Jamaica in 1896 to study tropical sea life, he contracted yellow fever and was the only survivor of six victims. When the fever left him, he found himself deaf, and this "made difficult the contact with people he so enjoyed" (Chace, 1947, p. 611). Clark went on to receive his doctorate from Johns Hopkins University. He spent two years at Amherst College as an instructor in biology and in 1899 he was appointed professor at Olivet College. In 1905 Alexander Agassiz invited him to join the staff of the Museum of Comparative Zoology, and five years later he was appointed curator of Echinoderms. In 1927, Clark became curator of marine invertebrates and associate professor of biology. As Chace (1947) has explained, "Clark's entire background fitted him for the teaching profession, but when deafness interfered with this calling he loved so well, he became an excellent museum curator with one of the best-arranged and richest collections of echinoderms in the world to his credit" (p. 612). Clark's collecting work brought him to Jamaica, Bermuda, Tobago, the coasts of North, Central, and South America, the Galapagos Islands, Australia, China, and Japan. He was particularly attracted to Australia and much of his work took place in that land. Twenty of his publications were on the distribution, variation, anatomy, and pterylography of birds. He was also interested in reptiles and amphibians. More than one hundred of his published reports focused on echinoderms. His *Catalogue of Sea-urchins in the British Museum*, published in 1925, contained about 8000 specimens belonging to nearly 400 species. Additional visits to Australia in 1929 and 1932 resulted in a work of even larger scope, *Echinoderms from Australia*. In what is perhaps Clark's greatest work, *The Echinoderm Fauna of Australia*, published in 1946, he brought together all that is known of that fauna, recent and fossil, so much of it contributed by himself. For this he received, on his deathbed in 1947, the Australian Clarke Memorial Medal.

In contrast to Clark's life, which was completely in the society of hearing people, the herpetologist Lewis H. Babbitt spent much of his time with other deaf people, especially school-aged children. Babbitt lost his hearing at the age of two and a half and attended the Clarke School for the Deaf in Northampton, Massachusetts and the American School for the Deaf in Hartford, Connecticut. He and his wife, Corrine, traveled thousands of miles each year in their search for reptilian and amphibian specimens for museum collections and displays. He authored dozens of bulletins and other works on herpetology. He was a collector of fauna, an honorary member of the New England Museum of Natural History, and a curator for the Worcester Natural History Society. One will find photographs of the deaf herpetologist in the September, 1939, issue of *New England Naturalist*, stalking the blue-tailed skink in Connecticut (Babbitt, 1939). He and his wife lectured at many schools. For regular schools, Corrine presented and Lewis handled the media and specimens. Lewis communicated in sign language when they visited schools for deaf children. Both at large assemblies and in individual classes, they brought numerous live scorpions, lizards, iguanas, and other reptiles and insects. Babbitt and his wife established a private collection called the Babbitt Trailside Museum at their two-hundred-year-old home in Petersham, Massachusetts. For years he edited the bulletin, *The Reptiles of Connecticut*. In 1937, he authored another bulletin, *Amphibia of Connecticut*. Specimens he and his wife caught were sold to scientific supply companies for school use, and to the Ross Allen Reptile Institute at Silver Springs, Florida. The venom from snakes, and also newts and other specimens, were collected for skin grafting, thyroid studies, and other medical research (Stansfield, 1947).

At the time World War I had ended deaf physicians were still a rarity. *The Volta Review* actually placed an advertisement in 1923 in search of one (*25*[4], p. 201). Still a profession punctuated with attitudinal resistance, the medical fieled imposed on both congenitally deaf and adventitiously deaf men and women attitudinal burdens too heavy for most to carry. Late-deafened physicians practicing medicine in the years immediately following World War I included Sir Charles Sherrington, always apologetic about his hearing problem (Brain, 1964), who held the office of president of the Royal Society in 1920 and was a member of the Order of Merit. Sherrington won the Nobel Prize for physiology or medicine in 1932 for his research on the nervous system. Thomas Milton Rivers, advised to forget medical studies because of his hearing loss, ignored this and went on to distinguish himself. He had Aran-Duchenne type of muscular atrophy, and a childhood mastoid operation left him without an eardrum in one ear. Rivers earned his medical degree at Johns Hopkins in 1915. He was elected a member of the National Academy of Sciences. The pathologist Eugene Lindsay Opie, distinguished for his pioneer studies, continued to practice medicine after the onset of his deafness. Another surgeon, Dr. Berger from Denmark, requested that his mastoid cells be opened, one of the earliest operations of this kind on record. His colleague was unfortunately not successful and, as Clendenning (1933) explained, the case acquired considerable fame when Berger's death from meningitis shortly after the operation appeared to have frightened surgeons from performing mastoid operations for years. Lucius W. Case was born of missionary parents in Parral, Mexico, and graduated in 1910 from Pomona College in Claremont, California, and from Ann Arbor Medical School in 1914. His deafness came on rapidly after his arrival at a hospital at Davao, Mindanao Island, in the Philippines (Davies, 1924, p. 122). Case had answered a call of the American Board of Foreign Missions for a physician and surgeon to take charge of the hospital and, although completely deafened, he did not write home about it for the three years he was there. During World War I, he returned to California, where he learned that his deafness was incurable. After further study in bacteriology and pathology, Case was offered a position in a laboratory in Los Angeles and eventually became head of the department.[4]

Glimpses into the lives of these late-deafened physicians add relatively little to our understanding of the barriers deaf persons faced in pursuit of training in medicine. At this point in history there were few prevocationally deaf people in pursuit of medical degrees. Physicians who were deafened *after* they had obtained their medical degrees were obviously fortunate in not having had to face admissions committees of the medical schools who would likely reject them. It was not until the mid-twentieth century that educational barriers were sufficiently overcome for the pursuit of a medical degree by any persons born deaf. Today there are still too few deaf men and women in the medical profession, and it continues to be difficult to effect widespread attitudinal change. Each new deaf medical student is largely left to face these attitude barriers alone.

Nevertheless, there were individuals who found ways to overcome the barriers of their day. Clyde S. Jones, profoundly deaf since the age of eighteen months, became the first deaf person to earn two doctoral degrees (doctor of public health and M.D.). Jones headed the Veterans Bureau Laboratory in Milwaukee, taught histology and pathology at the St. Louis College of Physicians and Surgeons, and was the city bacteriologist of East St. Louis for eight years (Burnham, 1924). He was a former pupil of the Illinois School for the Deaf, took advanced courses in chemistry and bacteriology, and earned a master's degree in bacteriology. His lectures at the St. Louis College of Physicians and Surgeons were primarily through writing on the blackboard. Jones became interested in experimental work on cholera and perfected a serum for diphtheria. In East St Louis, he was best remembered for the crusade he launched against impure milk, and with the cooperation of the state health authorities, Jones raised the standards of cleanliness in the dairy industry.

Along with Jones and Case, other deaf bacteriologists joined the war against disease long after the great breakthroughs of Pasteur, Lister, and Jenner, but their battles with bacteria, insects and public hygiene problems were rewarding. To be sure, the profoundly deaf French physician Charles Henri Nicolle ranks among the great contributors in this science for his discovery of the body louse as an agent in transmitting the highly infectious typhus fever. But while Nicolle was receiving his Nobel Prize, a half dozen deaf Americans were continuing their own scientific research to conquer disease. Ray and Arthur Wenger, both deaf since the age of two, opened Wenger Laboratories, where they conducted many studies. The Deaf community followed their scientific work closely, reporting on their many accomplishments in bacteriology through the years. "We do and should applaud such ambition for it helps the deaf as a class," one editor wrote in 1921. "When a deaf man rises he needs encouragement for still greater efforts" (*SW*, *34*[2], 1921, p. 76). At one point, the Utah Copper

Company contracted with the Wengers to assist them in connection with a $1.5 million lawsuit, one of a number of lawsuits in which they were asked to become involved. The firm was accused of damaging farms and health. The Wengers carried out a variety of tests and their studies were instrumental in the court's finding of no cause for action. Ironically, the same firm had turned the scientists down for employment a few years earlier—because of their deafness.

Anthony A. Hajna was born in Poland and immigrated with his parents to the United States. He lost his hearing when he was five years old, as a result of meningitis. Hajna became one of the nation's authoritative scientists in enteric bacteriology, authoring many papers and designing laboratory procedures for identifying epidemic-type forms of bacteria. In 1930, he graduated from Gallaudet College. He completed his master's degree in the School of Hygiene and Public Health in 1932 at Johns Hopkins University. In that same year he was appointed assistant bacteriologist in the central laboratory of the Maryland State Department of Health where he remained for seventeen years. For two years he was employed by the Vermont State Board of Health, training workers in the state laboratories, then at the U.S. Public Health Service in Cincinnati, Ohio, where he was involved in membrane filter studies. This work prepared him for his position at the Indiana State Board of Health. Here he spent many years researching techniques to accurately and quickly identify epidemic-type bacteria, especially those found in spoiled food and contaminated water. Hajna's skills as an epidemiologist were called upon numerous times. One case occurred when an outbreak of food poisoning sickened many nuns who had gathered together for a conference. The state board of health conducted numerous tests, but found nothing in the milk and food. Hajna examined the conditions which led to the food poisoning and discovered that one of the people used to prepare the food was an elderly woman carrying the typhoid bacterium (J. M. Smith, 1967). Hajna served on the editorial board of the *Laboratory Digest*. He published many reports on his scientific work. In 1965, he was honored with the State Achievement Award from the executive Audial Rehabilitation Society of Indiana for his "outstanding achievements despite the loss of hearing in early childhood" (*DA*, *18*[6], 1966, p. 17).

Another early bacteriologist was Averill J. Wiley, deaf since the age of fifteen. He was technical director of the Sulphite Pulp Manufacturers' Research League, responsible for reducing stream pollution caused by effluents discharged from paper mills, and the conversion of dissolved wood components in effluents to yield commercially valuable products. Wiley attended Whitworth College, then transferred to Washington State University at Pullman, majoring in bacteriology and public health, and received his B.S. degree in 1935 and his master's in 1936. From 1938 to 1941 he continued his graduate work in fermentation biochemistry and sanitary chemistry at the University of Wisconsin at Madison. Wiley was responsible for a staff of chemists and technicians. He authored several chapters for books and more than thirty technical papers in fermentation biochemistry. In addition, he held a number of patents (Horgen, 1963).

The unpublished autobiography of Harold J. Conn (1886–1975), located in the Cornell University archives, is one of the few detailed personal accounts

Harold J. Conn, American Bacteriologist

available on the effects deafness may have on a scientist's life. Conn's deafness began when he was about eleven years old; he later wrote of the vain hope of hearing again and how communication with him became a nuisance for others. Conn contributed much to the field of bacteriology, including the publication of several books on the subject, and he played a significant role in the founding of the Biological Stain Commission. In 1917, the Society of American Bacteriologists appointed Conn chairman of a committee to develop a descriptive chart of cultural characteristics of bacteria. For various reasons, the members of the committee were not very productive, and later, when the society requested that a complete manual of methods outlined by the chart be developed, Conn requested that a committee on "bacteriological technic" be established consisting only of members who were able to take an active role. "Making this suggestion was one of the hardest things I ever did," he explained, "and at the same time one of the most important steps I ever took. By that time my deafness had gone to a point where I could hear none of the business transacted, and I had never found any of the crude hearing aids of those days to be of help; so just finding a chance to make my motion was difficult."[5] His extensive work led to a number of popular books in his field, including *Biological Stains* (1925), *History of Staining* (1933), *Manual of Microbiological Methods* (1957), and *Staining Procedures* (1960).

Earlier (Chapter 2, "The Deaf Physician and the Barrier of Attitude") I presented a few extracts from a series of letters published in *The Lancet* in 1884, begun by "Constant Reader," who had expressed his concern about the onset of deafness and the effects it may have on his career as a physician. This was, coincidentally, the same year that another young man in France became profoundly deaf. Fortunately for the field of medicine, Charles Henri Nicolle (1866–1936) overcame attitudinal barriers. Nicolle went on to win the Nobel Prize for his development of the procedures for fighting typhus fever. He passed his examination for a medical residency when he was twenty-three years old. This accomplishment must have been challenging for him. His deafness had begun when he was eighteen and "created difficulties in his medical practice and in his professional contacts" (M. D. Grmek, 1978, p. 454). Grmek described the effects Nicolle's deafness had on his interaction with others: "An introvert by nature, Nicolle dreaded social gatherings because his deafness excluded him from the conversation. He divided his time among scientific research, writing, and his family" (p. 454). Lot (1961) explained that Nicolle's deafness gave him a false air of inattention. He would often excuse himself mischievously, shrugging off his deafness with a remark about the absent mindedness of many learned scholars. Nicolle believed that his "faulty ears" had provided him with a turning point in his life. He felt that had he not become deaf, he may not have turned to the laboratory. Nicolle confides in his "Letter to the Deaf" the precious stimulation he had found in his hearing loss. He had given up medical practice as a result of his deafness, but the loss of hearing had rescued him, made him stronger, and, in effect, directed him to great things. Duhammel described how Nicolle was separated from the world by an insurmountable barrier, "the sort of trouble that discourages many people, but which had placed him in extraordinary conditions favorable to meditation and invention" (Lot, 1961, pp. 23–24). In

The Biology of Invention (1932) Nicolle used the metaphor of a man wandering about with no direction and coming to a precipice. He described how deafness had saved him: "To my disability I attribute the fact that I had to leap to the place I now find myself. I had no choice. I had to be stronger than misfortune, or succumb to it" (Lot, 1961, p. 23).

Nicolle was the first deaf Nobel laureate; the second was Sir John Warcup Cornforth in 1975. Three years before Nicolle won, however, Henrietta Swan Leavitt was nominated by a Swedish mathematician, Professor Mittag-Leffler, for her work in astronomy, but unknown to him, Leavitt had died of cancer. Oliver Heaviside was also nominated in 1912. In a letter from the physicist Vilhelm Bjerknes to G.F.C. Searle, Heaviside's most loyal friend, the Norwegian professor wrote, "I proposed Heaviside for the Nobel Prize. But, alas, it was 100 years too early" (Searle, 1950, p. 8).

Several other breakthroughs should be mentioned. Tracy Hinkley became the first deaf person to hold a position in the field of physics in modern times. He worked with uranium ore, extracting radium at a plant in Montrose County, Colorado. He also studied radon in cancer research at the City Hospital in St. Louis. "Hinkley did not become a deaf-mute [sic] in the service of science," wrote the editor of *The Catholic Deaf-Mute*, "rather, science has proved his escape from his condition, as he was afflicted at birth" (*SW*, *39*[9], 1927, p. 339). And David H. Wilson may have been the first American born profoundly deaf to earn a Ph.D. Wilson was born in Philadelphia and attended the Wright Oral School. His father was Dr. William P. Wilson, director of the School of Biology of the University of Pennsylvania and a founder of the Philadelphia commercial museums. His mother, Dr. Lucy Langdon Williams Wilson, has been appropriately credited by the Deaf community for her loving support that certainly contributed to David's success. She was head of the Biology Department of the Philadelphia Normal School. Suffragists and members of women's clubs rallied behind her when she applied for principalship of the Southern High School for girls. These same qualities helped her when she discovered her only son, David, was deaf: "Instinct as well as observation convinced the scholar-mother that her son was a child of extraordinary mental calibre" (*SW*, *28*[2], 1915, p. 37). In her teaching, she won renown for her advocacy of individualization of instruction, and she applied this interest to the instruction of her son by inviting the head of the Government School for Mutes in Vienna to the United States for one year to advise on methods of educating David. By the age of nine, David was considered a "wonder." He mastered the study of astronomy so well that scientists permitted him to use the Philadelphia Observatory. David entered Harvard University, where he earned a B.S. degree *cum laude*. He completed his Ph.D. at Harvard in 1931, majoring in philosophy, but his interests in the interrelationships of the sciences and humanities led to various scientific reports.

The Study of Cultures

Through the years since the Deaf community was established in the United States in the nineteenth century, many deaf people have written books on their personal experiences. In a sense, these books are ethnographic documents, autobiographical accounts of the ways in which partial or complete deafness can shape an individual's perspectives, motivations, happiness, and success. Anthropological writings of deaf men and women have more recently focused on a systematic study of the "Deaf culture" and "Deaf community." Carol Padden and Tom Humphries are two of the more popular writers in this area. In their book *Deaf in America: Voices from a culture* (1988), they examine the way members of the Deaf community live and communicate and how this compares and contrasts with the way hearing people live and communicate. The further we go back into history, the more we learn about the broader interests of deaf ethnographers in America. In Chapter 2, I briefly described Robert J. Farquharson's ethnographic work in regard to the Native American Mound Builders. This was the subject on which he had presented his lecture at the American Association for the Advancement of Science. Farquharson was eventually rewarded for his anthropological work when he was appointed a member of the French Société Ethnographique in 1880. The membership was something which he always considered one of the highest honors awarded him. In 1880, he was also made a fellow of the American Association for the Advancement of Science. Two other examples are George E. Hyde (1882–1968) and Dale L. Morgan (1914–1971), both of whom contributed significantly to understanding the cultural aspects of the American West. Morgan, deafened at the age of fourteen, authored such historical writings as *Jedediah Smith and the Opening of the West* (1953) and *The West of William H. Ashley* (1963). Hyde was a Plains Indian historian and assistant to the anthropologist George Bird Grinnell. As John Panara (1987) has explained, Hyde was one of the first to obtain a Native American version of the past, "and it also gave him the opportunity to develop both a balanced sense of history and a profound understanding of Indian identity" (p. 79). Hyde was profoundly deaf and partially blind.

Hyde's assistance to Grinnell led to a very productive career of his own. This was not the case for Sebastian Putnam, the deaf brother of the anthropologist Tracy Lowell Putnam. Sebastian helped his brother to study the pygmy tribes of the Belgian Congo in the 1930s and 1940s. Although he never became more than a field assistant, he did manage his brother's camp during a period in 1928. He got along well with the African bearers, learning from them a Mangbetu betting game which he then taught his brother (who subsequently published on the subject). The deaf man's popularity among the Africans, explained Professor Joan Mark from the Peabody Museum of Archaeology and Ethnology at Harvard University, came partly from his ability to communicate easily without using an oral language.[6]

Without question, the greatest anthropologist to overcome the challenges of deafness was Ruth Fulton Benedict (1887–1948). Despite the hindrances it may have introduced and the emotional turmoil it caused her throughout her life, deafness resulting from an attack of measles as a child did not prevent Benedict from great achievement. She went on to become one of the first female social scientists of renown. As for how her deafness may have affected her cultural observations, Donald Fleming (1971) has written that "she was forced to rely entirely upon English-speaking informants and interpreters; with her deafness, she would have been hard put to learn a primitive language by ear" (p. 131). In 1922, she worked with the Serrano Indians at the Morongo Indian Reservation in the San Gorgonio Pass in California. The primary Serrano informant, Flora, affectionately referred to Benedict as "a deaf" (Caffrey, 1989, p. 109). While with the Serrano Indians, she also worked with Alfred Kroeber, another former student of Franz Boas. Kroeber not only shared similar anthropological interests with Benedict, but the experience of deafness as well. For the previous seven years he had had symptoms of Ménière's syndrome, including vertigo and hearing loss, which left him totally deaf in his left ear (Kroeber, 1970). As Caffrey (1989) has written: "Who better could have directed Ruth Benedict in the field than a man who understood her problems with an insider's view and could therefore suggest effective ways to overcome her hearing problem and be productive? In return she may have helped him understand life with a hearing handicap and how to make the best of it" (p. 104). Kroeber's meticulous work with Ishi, a Yahi, was highlighted in the film, *The Last of His Tribe*.

In *Ruth Benedict: A Humanist in Anthropology* (1974), Margaret Mead described Benedict's turbulent childhood, her struggles in school and marriage, and how her deafness led her to the more visual aspects of culture, including poetry, architecture, and painting. Her experiments with hearing aids proved futile, and she found it frustrating to communicate with her students at Columbia University. Yet Benedict overcame these barriers and became a leading American anthropologist. She is best known for her book *Patterns of Culture* (1934), widely used as an introduction to anthropology for decades after its publication. She also published *Race: Science and Politics* (1940) and *The Chrysanthemum and the Sword: Patterns of Japanese Culture* (1946); the latter work which brought a large grant to Columbia University for a program entitled "Research in Contemporary Cultures" which she directed.

Astronomy and Space Science

In addition to Annie Jump Cannon and Henrietta Swan Leavitt, who, as discussed in Chapter 2, discovered many new binary stars as well as other celestial phenomena such as comets and novae while working at the Harvard College Observatory, several other deaf persons contributed to astronomy during this period. One was Olaf Hassel (1898–1972) of Norway. If there had not been a shortage of oil during World War I, Hassel may never have turned his interests to the heavens. As a consequence of the shortage, school opened late in 1914; and during his free time, Hassel studied the constellations, helped along by an older brother who was more experienced. While his family and friends passed the dark hours in conversation or listening to music, Hassel, now a teenager, began studying the stars with a crude telescope. A lens from his mother's spectacles

Olaf Hassel, Norwegian Astronomer

Comet Jurlof-Achmarof-Hassel (1939)

served as the objective; a magnifying lens as the eyepiece. It was a remarkable similarity to the way John Goodricke had begun his work more than 130 years earlier. In his autobiographical notes shared with me by his niece, Mrs. Reidun Guldal, Hassel wrote that "some people have written and spoken about the bright years of my childhood. But for me, without hearing, my childhood years were the worst and darkest of my life. I soon noticed that I was different from other children and adults who could hear and talk, and I was of course disconsolate with my heavy destiny."[7]

After attending the Christiania Public School for the Deaf in Oslo, Hassel worked at the Meteorological Institution for twenty-seven years. He was awarded several grants from the Nansen Foundation, including one for a solar eclipse expedition. He assisted Fredrik Carl Størmer for many years in investigating the polar aurora and other phenomena. In January, 1938, during the magnificent polar aurora from which light was visible in France and as far away as North Africa, Hassel and Størmer made 400 photographs. The interesting luminosities in the atmosphere were rewarding experiences during the long hours in which they studied the auroral displays. Hassel was honored for having discovered "Comet Jurlof-Achmarof-Hassel" in 1939 and "Hercules Nova 1960 (Hassel)." He received the American Association of Variable Star Observers Fritjof Nansen Award and a gold medal from the king of Norway, and he was elected an honorary member of the Norwegian Astronomical Society in 1969, three years before his death.

Around the same time Hassel began his work, Robert Grant Aitken (1864–1951), an American who was progressively deafened since childhood, published his book *The Binary Stars* (1918). This book was based on Aitken's own discoveries of more than 3000 binaries. Aitken's work made reference to the seminal studies of Cannon and Leavitt, both of whom were still working at the Harvard College Observatory. According to W. H. Van Den Bos (1958), Aitken's deafness was "an affliction of long standing" (p. 6). A hearing aid helped him to follow a lecture, but without it he was functionally deaf. He was once struck by an automobile on a street in Berkeley when he failed to hear its approach. Fortunately, he recovered from the accident, and the astronomical community continued to benefit from his excellent writing. Aitken's hearing loss increased when he was about twenty years old. A copy of his "Record of Family Traits" filled out for the Eugenics Record Office was sent to me by Janice F. Goldblum of the National Academy of Sciences. Aitken attributed his "defective hearing" to catarrhal trouble (probably catarrhal otitis media, an early name for a type of middle ear loss). He may have also had some hearing loss as a result of several severe attacks of pneumonia during his childhood which prevented him from attending school until he was nine years old. In addition, Aitken had scarlet fever when he was young.

In W. H. Van Den Bos's obituary in *Monthly Notices of the Royal Astronomical Society*, there is an anecdote about Aitken with which nearly every deaf person can identify. It happened at the first International Astronomical Union Assembly in Rome in 1922. A foreign astronomer wished to discuss an aspect of double-star studies with Aitken, and, faced with Aitken's deafness as well as the necessity of speaking in English, he spoke at the top of his voice.

As Van Den Bos (1952) explained, the attempt was not a conspicuous success. Giving up in despair, the astronomer whispered, "What a blessing for Astronomy that his eyes are so much better than his ears!" (p. 273). Aitken, watching his colleague's lips, responded in agreement and Van Den Bos related, "As happens so often with the deaf—[Aitken] had missed most of the shouting, but understood the whisper perfectly" (p. 273).

Aitken published several hundred articles and reports on binary stars, satellites of planets, and comets. His greatest work was *New General Catalogue of Double Stars Within 120° of the North Pole* (1932). He received awards from scientific societies in both Europe and the United States and held the office of president of the Astronomical Society of the Pacific.

In Russia, Konstantin Eduardovich Tsiolkovsky (1857–1935), deafened at the age of twelve, became a national hero for his detailed work in rocketry. His childhood reminiscences were dramatic:

> I often behaved awkwardly among other children my age, and among people generally. Naturally, my deafness made me ridiculous. It estranged me from others and compelled me, out of boredom, to read, concentrate, and daydream. Because of my deafness, every minute of my life that I spent with other people was torture. I felt I was isolated, humiliated—an outcast. This caused me to withdraw deep within myself, to pursue great goals so as to deserve the approval of others and not to be despised. (Riabchikov, 1971, p. 92)

Tsiolkovsky believed his life was lacking in human contacts. In an effort to recall the sad years when he was between ten and fourteen years old, he could not think of a single event: "My deafness made my life uninteresting, for it deprived me of companionship, of the possibility to hear people around me and to imitate them" (Kosmodemyansky, 1956, p. 9). In order to help himself hear human voices, he built a set of tin funnels. He remained isolated from those who did not seem to understand his interests in science, and the ear trumpets only added to his eccentricity. His hair remained uncut only because he did not wish to take the time to bother with it, and his clothing was often stained and eaten by acids from his chemical experiments. It is no wonder that local villagers followed the scientific work of the deaf "crank" with curiosity and, at times, agitation. In the midst of all of this, Tsiolkovsky demonstrated a sense of humor. Visitors would often receive surprise electric shocks from his "octopus," a device which sent sparks from their bodies and caused their hair to stand on end.

Tsiolkovsky was the first of the great triumverate of pioneers in rocketry. He paved the road for the work of Robert Goddard and Hermann Oberth. In *Free Space* (1883), one of his hundreds of published works, Tsiolkovsky laid down the fundamental design of a rocket-powered spaceship. His thesis that the freedom of space would transform human life to a happier state grew from a firm conviction that this is what the future holds for humanity. The deaf prophet of space travel wrote: "Man will not always stay on earth; the pursuit of light and space will lead him to penetrate the bounds of the atmosphere, timidly at first, but in the end to conquer the whole of solar space" (Tsiolkovsky, 1960, p. 95).

In 1933, the Soviets tested a liquid-propelled rocket which represented the fulfillment of Tsiolkovsky's lifelong dreams. Oberth wrote to him in 1935, "You have ignited the flame, and we shall not permit it to be extinguished; we shall make every effort so that the greatest dream of mankind might be fulfilled" (Grigorian, 1976, p. 484). Sputnik and Sergei P. Kovolyov's design of the spaceship that carried the first cosmonaut, Yuri Gagarin, into space were largely influenced by Tsiolkovsky's writings. Had events turned out as planned, the launching of Sputnik would have honored the deaf rocket pioneer exactly 100 years after his birth. In 1935 a state funeral was held for Tsiolkovsky, the "Father of Astronautics."

THE SECOND WORLD WAR

When World War II began, many deaf people moved to Newport News, Virginia, to help with the defense industries, including the production of aircraft. Thousands of others found employment in the Akron, Ohio, area, where many war industries were based. Among them were Fred Schreiber and Ed Carney, both of whom became well-known leaders in the Deaf community. Schreiber, with his recently earned degree in chemistry, helped produce materials for Firestone Company for the war. He later served as NAD executive director. Carney left Gallaudet College at the start of the war and volunteered to help make gun bases. After completing his education at Gallaudet, he later served as the executive director of the Council of Organizations Serving the Deaf.

During the war, the NAD raised almost $8,000 for the purchase of vehicles for the Red Cross "Clubmobiles" for use in Europe. Deaf scientists also joined the war efforts. Einer Rosenkjar, a civil engineer, secured employment with the United States Engineering Office in Los Angeles. In this division of the War Department, he was called upon to take responsibility for the design of bridges along the Pan-American route in the Canal Zone. For these efforts he was promoted to associate engineer. Although hearing was considered essential for higher classifications, Rosenkjar was later promoted to senior structural engineer associate and for many years was involved with the design of freeways, parkways, and bridges in Los Angeles. Emil Gunnar Rath was offered a position supervising statistical clerks in the United States Air Force Headquarters after the Pearl Harbor attack. Rath advanced to analytical statistician and became the first civilian to receive the USAF Superior Accomplishment Award in the Directorate of Statistical Control. The analytical chemist Edwin E. Maczkowske, with the National Bureau of Standards since 1927, assisted with the research for the Manhattan Project, and Leon Auerbach was involved with research on radar electronics at the Massachusetts Institute of Technology until 1944. Robert H. Weitbrecht, a physicist, developed a specialized radio system for precision orbiting of an aircraft around a ground reference point while at the Radiation Laboratory at the University of California. He, too, worked with the Manhattan Project during the war, as well as with the Aeromedical Laboratory. One of his projects was an investigation of oxygen safety systems for aircrews. He also studied automatic parachute releases and was a scientific observer in a flight across the country, returning on a B–24 "Liberator" airplane. After the war Weitbrecht joined the U. S. Naval Air Missile Test Center at Point Mugu, California and spent four years developing electronic timing systems for the optical and radar missle range instrumentation. His research led to the development of modern digital frequency counters, and this work prepared him well for his subsequent patent for the acoustic coupler, which brought the telephone into deaf people's lives in the 1960s.

The recollections of deaf scientists about the war were both humorous and traumatic. One story involved the "foreign" quality which is often assigned to a deaf person's speech. Some deaf people are told they sound "French," while the speech of others has been described in relation to different languages. Robert E. Stansfield (1947) described an incident in which the herpetologist Lewis H. Babbitt's deafness lent color to an already colorful life:

> During [World War II], while the Babbitts were motoring near the border on their way to Canada, the man's [guttural] voice attracted the suspicion of a hotel bellhop who described him as an escaped German prisoner of war. Still farther on, at Owl's Head, a woman drugstore clerk, overhearing the pair talking about "pitcher plants," which she understood as "plans," tipped Federal men that a German agent was loose. Customs men stopped the Babbitt car and questioned the naturalists interminably. Attempts to question Mr. Babbitt proved fruitless. "Can't he speak anything but German?" they asked his wife. "Certainly," she answered. "He speaks English and French." (p. 2)

R. M. Eagar, who was deafened by spinal meningitis in 1941 at the age of twenty while an officer-cadet in Britain's Royal Engineers, continued his studies at Oxford University and completed a degree in geology with honors in 1942. After earning his Ph.D. in 1944, he continued his geological studies, particularly on shell variation in coal measures. In a short autobiographical sketch, Eagar described the danger associated with doing scientific research during World War II. Fieldwork was especially difficult, and on several occasions Eagar was arrested as a spy by the police or Home Guard:

> I was once in imminent peril of my life through failure to lip-read a sentry correctly. With his sanction, as I believed, I was happily digging away on the embankment of a reservoir, when I turned to see two very hostile soldiers bearing down on me with fixed bayonets. They had been shouting for some time and had evidently decided that only thus would they remove me from the target of a large gun firing shrapnel! The noise must have been terrific. (Eagar, 1947, p. 327)

Emmanuel Meeron, born in Warsaw, Poland, became totally deaf from meningitis at the age of four and a half and was a teenager when the war began. He escaped with his family to Italy, then on to Israel, fortunately avoiding the fate that befell other deaf Jews at the hands of the Nazis. Meeron enrolled in Hebrew University in 1945. While a student there, he published popular articles on various scientific subjects. In November, 1947, he joined the service when disturbances broke out in Israel, but he was barred from combat duty.

Wyn Owston, British Medical Doctor

Nevertheless, he was wounded twice. In January, 1950, Meeron was offered a post with the Government Research Laboratories, where he conducted investigations in organic chemistry at the Weizmann Institute. Later, he completed his master's degree at Hebrew University.

Among the few deaf scientists with doctoral degrees in the 1940s, three had stories to tell of the alarming journeys across the perilous waters during the early years of the war. Wyn Owston, then going by her maiden name Ethel Sharrard, ignored the advice of friends and family, who feared for her safety as England declared war on Germany. She returned to her homeland to complete medical school. But when Owston applied to the medical course of study at Sheffield University, she faced an unwilling and rude dean, even though both her parents were physicians and supported her application. Deaf and unable to speak clearly, she was not regarded by the dean as a good candidate for medical school. Through perseverance, she eventually succeeded and became the first deaf woman to earn a medical degree.

During the war Owston assisted in operations, supervising treatment, often for eighteen hours a day. She graduated in 1943, declined a position as house-physician in Birmingham and a research post, and focused on general medicine and surgery with the soldiers returning from Normandy and other war fronts. After World War II, she moved to Lincoln, where she took up obstetrics and worked in family planning clinics.

Owston's journey was within months of Donald L. Ballantyne's voyage to America from Hong Kong to continue his own studies. During the trip to the Archmere Academy, Ballantyne was on the SS *President Hoover* from Hong Kong to Shanghai when the vessel was bombed by the Chinese Air Force, which had mistaken it for one of the Japanese fleet. When a bomb hit the adjacent cabin on the boat deck, the profoundly deaf Ballantyne opened his door, thinking someone had pounded very hard. Ballantyne completed his doctoral work at Princeton and became a leading authority on microsurgical techniques. Rejected by medical schools because of his deafness, he went on to instruct physicians and surgeons in microvascular surgical techniques at the Institute of Reconstructive Surgery at New York University Medical Center.

Not long after Ballantyne's adventurous voyage came Tilly Edinger's trip across the Atlantic. Edinger was a "displaced scientist" of Jewish heritage whose brother had died in a concentration camp and who had escaped oppression in her homeland, first across the English Channel, then later to America. Fortunately, Edinger also arrived safely and eventually helped to found the science of neuropaleontology, the study of fossil brain cavities. She published more than 100 books and scientific articles and more than 1200 reviews and abstracts. Most of her research in the United States was at the Museum of Comparative Zoology in Boston. In 1950, Edinger won a fellowship from the American Association of University Women to update her comprehensive survey of the brains of extinct animals she had earlier published in 1929. She was struck by a car on May 26, 1967, and died soon after.

The general contributions of people with disabilities in industry during World War II, particularly war-related industry, helped to reconceptualize rehabilitation in the United States in the years to follow. In this sense, both world

Tilly Edinger, American Paleontologist

wars were turning points which led to important legislative acts that greatly enhanced opportunity for the deaf professional as well.

5

The Modern Era: Revolutions and Breakthroughs

The years immediately following the Second World War saw the National Association of the Deaf rebuild itself. State associations were set up, and this helped to make the continued fight against paternalism in regard to driving privileges, preferential tax treatment, and other issues more effective. Overall, college enrollment of deaf men and women across the United States remained at a figure estimated to be less than four hundred and deaf graduates of college programs likely never exceeded fifty or sixty per year (Moores, 1987, p. 403). Identification of the scientific contributions of deaf men and women continued to depend largely on the editorial prerogatives of the staffs of the magazines in the Deaf community. Occasional reports appeared which highlighted the recognition granted deaf persons in the form of honors and prizes by professional societies. In 1950, for example, the deaf biologist Robert J. Lillie (U. S. Department of Agriculture), won an award for the best paper of the year in poultry science, and the Firestone chemist Robert O. Lankenau was honored in 1981 for distinguished service by the Akron Council of Engineering and Scientific Societies. Jay Justin Basch (U. S. Department of Agriculture), Robert L. Bates (U.S. Navy Bureau of Aeronautics), and Edwin J. Parks (National Bureau of Standards) were all recognized for their exemplary service to their agencies through the program now titled the "Presidential Awards for Outstanding Federal Employees with Disabilities." Basch, born deaf, earned his Ph.D. from Temple University. He was recognized in 1970 for his extensive biochemical and biophysical research on concentrated milk products. Bates, a mathematician, was recognized in 1974. He was honored for his work in programming the Navy's Missle Performance Analysis, Rocket Trajectory Range, Atomic Shock Arrival Effects on Aircraft, and the U-2 Climb Path by Calculus of Variation. Parks, who became deaf at the age of fourteen and earned his Ph.D. in chemistry from American University, began his career as a research chemist in chemical and biodegradation processes at the NBS in 1957. He won the Presidential Award during the National Year of Disabled Persons (1982).

The dearth of biographies and statistical analyses of job market data substantiates the underrepresentation of deaf men and women in science. Persistent inquiry in the Deaf community has helped to provide a basic portrait of contributions in the modern period. In the post–World War II years, mathematics and physics began to attract deaf persons. In New Mexico, Donald H. Bradford, profoundly deaf since the age of seven from spinal meningitis, and a graduate of Gallaudet College, assisted computer pioneer Nick Metropolis in the Los Alamos National Laboratory's Theoretical Division with the development of one of the first computers. He later worked on the staff developing MADCAP, the first computer program. Simon Carmel, born profoundly deaf, and the only physics major at Gallaudet College in the early 1960s, worked for twenty years at the National Bureau of Standards in Washington, D.C., as a physicist, first conducting experimental work with weights and measures, then as an X-ray diffraction crystallographer. At the Stanford Research Institute, Robert H. Weitbrecht designed and constructed a precision camera for use with the Lick ninety-two-centimeter refractor telescope, which featured photoelectric guiding, and which provided high-quality photographs of star fields, nebulae, the moon, asteroids, and planets.[1] Ralph Guertin was probably the first deaf person to earn a doctorate in physics. He received his master's degree (1963) and his Ph.D. (1969) from Yale University. Ignoring the advice of others that "a deaf person should not be teaching" (Gavin, 1975, p. 721), he took a position as a professor at Rice University for six years. Guertin later held the office of vice president for geophysical development with O'Connor Research, Inc. Donald J. Kidd, one of the first congenitally deaf persons to earn a doctoral degree in science in Canada, conducted research at the University of Alberta. He took positions as a research geologist with the Geophysical Engineering and Surveys Company in Ontario and as a consultant geologist for the Arctic Institute of North America.[2]

The federal government continued to be a primary employer of many deaf scientists. Between 1950 and 1963, for example, twenty-one out of forty-six Gallaudet graduates in science-related fields were employed by federal agencies. For the years 1962 and 1963, ten out of eleven obtained positions in the U.S. Army Corps, Army Map Service, Department of Agriculture, Coast Guard, Geodetic Service, or the Department of Health, Education, and Welfare.[3] Research chemists such as James L. Casterline and Thomas D. Doyle continue to distinguish themselves today in the Food and Drug Administration.

Several accomplishments by deaf scientists in the modern period have been notable. Frank Hochman became the first congenitally deaf American to earn a medical degree.[4] Hochman was chief medical technologist at Somerset Hospital in New Jersey in 1972 when he decided to enter medical school. After encountering many negative attitudes in his interviews, he was accepted at Rutgers Medical School at the age of thirty-seven. He earned his M.D. in 1976 and set up family practice in the East Bay area of California. In 1975, England's John Warcup Cornforth won a Nobel Prize for his research in cholesterol chemistry. Deaf since the age of ten, he began with a small chemistry laboratory at home, developed an intense interest in the subject, and went on to conduct research on the chemical structure of penicillin, the synthesis of steroids, and, eventually, the use of radioactive isotopes to study the formation of the molecule of cholesterol from the acetic acid molecule. David James earned his Ph.D. in mathematics at the University of Chicago in 1977 to become one of a very few African-American deaf persons to enter a science-related profession.

James, profoundly deafened at the age of four, was one of only six African-Americans in his high school graduating class. He procured a teaching assignment at Indiana University, then became an associate professor of mathematics at Howard University in Washington, D. C., where he has conducted research on differential topology and computer modeling.[5]

Technology continued to play an important supportive role, allowing people with any age of onset of deafness to participate in scientific research. Dr. Helen Brooke Taussig (1898–1986), who pioneered with Dr. Alfred Blalock in the "blue baby" operation in 1944, was one person who so benefited. The famous operation involved the insertion of a blood vessel between the subclavian and pulmonary arteries and quickly changed the infant from blue to pink. In the 1940s, Taussig, who experienced progressive deafness, used a stethoscope attached to a large black box which amplified the heart sounds. With this she was able to continue her outstanding research in congenital heart malformations.[6]

THE BARRIERS CONTINUE

At first glance, the breakthroughs might seem a beacon of change in attitudes, but the decades following World War II were a period of continued struggle. Alan B. Crammatte's survey of deaf persons in professional occupations provided, through self-reports, some detailed examples of attitude barriers faced by deaf scientists. Crammatte reported that nineteen different agencies employed twenty-four of the eighty-seven deaf respondents in his study. Of the forty-six deaf scientists (twenty-three of them chemists) and mathematicians responding to his survey, all had remained in occupations identical or related to their major fields of concentration during their undergraduate years. Only one of the respondents, a mathematician, was female, and about six out of ten in science-related occupations were either born deaf or deafened before the age of six. The deaf scientists responding to the survey believed that the outstanding characteristic that made their occupation "suitable" for deaf persons was the "absence of need for frequent oral communication" (p. 77). They described the attitudes they faced in being hired and the strategies they used to break down the barriers to secure careers in the field of science. One deaf chemist wrote approximately 750 letters of application and was continually frustrated with the reluctance of employers to hire him. Another respondent felt the resistance he experienced was "definitely antagonistic." A deaf statistican and section chief with the Railroad Retirement Board described this dilemma further:

> The attitude of an appointing officer (even one with the best will) toward a deaf person . . . cannot help but be colored by whatever initial difficulties he has in interviewing a deaf person for a job. He is likely to be unfavorably impressed if the interview is made more difficult and awkward because of the failure of the appointing officer and the applicant to understand each other. . . . Moreover, no matter how well a deaf person can speak, even by the standards of the non-deaf, an appointing officer will have difficulty understanding what is being said to him if he has allowed the knowledge that it is a deaf person speaking to "condition" him against understanding. (Crammatte, 1968, p. 65)

Sixty-four percent of Crammatte's survey respondents admitted to having help in securing positions through friends and family. Comments revealed how difficult it was for them to be hired into positions commensurate with their skills. "They just didn't want to have anything to do with handicapped people," wrote one scientist, and these employers "thought it would be too dangerous because I would have to go out into the foundry" (p. 68). Another respondent complained, "They told me that they were not against deaf chemists but they claimed that they had an insurance clause in their contract that was against hiring deaf people" (p. 69). "There was a chief chemist . . . that wanted to hire me but the medical department said they definitely wouldn't," wrote a third deaf professional (p. 69). Ironically, studies of deaf workers in several of the largest industries in the nation have shown their records to be superior to those of their hearing coworkers (Menchel & Ritter, 1984).

Most women professionals continued to enter fields other than science. Barbara C. Nies, deafened at the age of six, graduated from Wilson College in Chambersburg, Pennsylvania. She was hired at the Columbia University College of Physicians and Surgeons, where she spent eight years as a research assistant, including work for the physicians who developed the Rh vaccine "Rhogam." Nies left her work in medical research when she decided on a career in teaching deaf children.

Where were the deaf women college graduates? Nancy Kelly-Jones (1974) reported on a three-year sample of female Gallaudet graduates for the years 1971–1973. Out of fifty-eight women receiving degrees from the class of 1971, twelve became homemakers, eighteen took clerical positions with the government, and twenty-three were school counselors, supervisors, teachers, and aides. Five entered graduate school for continued study. From the class of 1972, eighteen were reported in graduate school, fourteen as homemakers, nine in clerical positions and six as school counselors, supervisors, teachers, and aides. She found sixteen in graduate schools from the class of 1973, while twelve were homemakers, five in clerical positions, and seven in positions as school counselors, supervisors, teachers, and aides. The fields of specialization of the deaf women in graduate schools are not reported. Science was not a popular choice. Even in 1987, only about thirty of the 1600 students at Gallaudet were science majors, and only half of this number were women (Mandula, 1987).

MacLeod-Gallinger (1992) presented data which illustrate the continued underrepresentation of deaf women in science. In a study of nearly 5000 deaf high school graduates, she reported that 4.9 percent of the deaf women chose instructional programs in allied health (medical laboratory, medical records, dental technology, nursing, and other programs associated with health care and medicine) as compared to 15.6 percent of hearing women in the national data provided by the U. S. Department of Education in 1988. In engineering 0.0 percent of the deaf women selected courses of study (compared to 1.5 percent of hearing women), and in engineering technologies, 0.1 percent of the deaf women enrolled in programs (compared to 0.8 percent of hearing women), prompting MacLeod-Gallinger to conclude that "very few had earned degrees in engineering, physical sciences, health sciences (includes medicine), communications, protective services, public affairs, and social sciences. Deafness most certainly factored into these exceptions, whether they resulted from real or perceived barriers" (p. 319).[7]

The real barriers encountered by deaf women scientists led Nansie Sharpless to join the Foundation for Science and Disability (at that time known as the Foundation for Science and the Handicapped) in the 1970s and advocate change. Sharpless, profoundly deafened at the age of fourteen, faced pessimistic attitudes about her pursuit of a doctoral degree in science even among her friends and family. She completed her degree with a 4.0 grade point average and proved the fears about overqualification unfounded. At Albert Einstein College, she directed a biogenic amine assay laboratory to study chemicals involved with brain functioning. She was a strong advocate for science careers for deaf women and presented lectures on this subject until her death in 1988.

The dilemma of attitude has continued to be especially problematic for both deaf men and women interested in the medical sciences. With a dream of becoming the first born-deaf physician in the United States, Michael Weiner applied to seven medical schools in Pennsylvania and was turned down by each of them. "It was the worst time of my life," he explained. "For the first time I really began to have doubts about myself and what I could accomplish" (*Philadelphia Evening Bulletin*, September 26, 1975). Among the reasons for rejecting him from medical school were that he would not be able to use the stethoscope and that the use of surgical masks would prevent him from communicating effectively. Weiner went on to earn a medical degree and set up a successful practice in northwest Philadelphia. James C. Marsters faced the same attitude when he was turned down by all the dental schools to which he applied in the late 1940s. Even though he scored impressively on the entrance examination for the New York University Dental School, his interviewers showed no interest in him. As with Weiner, persistence paid off and Marsters was finally admitted to New York University. After graduation, he set up practice with an orthodontic specialty in Pasadena, where he served on the California Board of Dental Examiners Committee.[8]

Still another example was Judith Ann Pachciarz, whose battle to enter medical school lasted seventeen years. She was deafened at the age of two and a half years by encephalomeningitis. In 1964, she graduated from the University of Illinois with a major in microbiology and zoology and a minor in genetics. That year she applied to many medical schools and was turned down by every one. Her application included an addendum: "My intent is to complete the full medical curriculum and all clinical rotations at the least cost and inconvenience to the school of medicine without compromising in any way the quality of patient care but, rather, enhancing that quality as perceived by the patient" (Moore, 1987, p. 5). After she completed her M.S. and Ph.D. at St. Louis University, Pachciarz took postdoctorate courses at the University of Florida in immunobiology and pathology. Florida's medical school turned her down even though she had been teaching medical and dental students there. In 1974, she took a position in the Department of Veterinary Science as an assistant professor at the University of Kentucky. By 1976, she had behind her fourteen years of applications to medical schools and sixteen rejections. She refused to give up. Each disappointment strengthened her determination. Her attorneys approached the University of Florida College of Medicine to negotiate with them on her behalf, citing spurious reasons denying her admission which they believed violated both state and federal law. Pachciarz viewed her struggle as not only a case of clarifying the rights of an individual with a disability, but one with national ramifications. Finally, in 1983, she was accepted at the Louisville School of Medicine and fulfilled her dream, becoming the first deaf woman in the United States to hold both M.D. and Ph.D. degrees. When she practiced medicine, she improvised, using an oscilloscope attached to a stethoscope. She supervised a staff of about two hundred people, including forty physicians, at the World Games for the Deaf in July, 1985, where 2,100 deaf athletes competed.

Frances Davis, a licensed practical nurse, took her case to the Supreme Court in 1979. During litigation, in opposing her effort to gain admission to a nursing program to be trained as a registered nurse, the counsel for Southeastern Community College argued that Davis could not safely participate in a normal clinical training program owing to her disability. The United States Supreme Court expressed a unanimous view that the decision of the training program was not unlawful discrimination. As DuBow (1987) explained in his detailed summary of this case, the Supreme Court concluded that "the ability to understand speech without reliance on lipreading is necessary for patient safety during the clinical phase of the program" (p. 414). The justices felt that the modification needed in the clinical program to accommodate Davis was greater than Section 504 of the Rehabilitation Act of 1973 required, and that the admission standards of the Southeastern Community College program would have to be lowered, again something not required by Section 504.

The decision angered advocates for scientists with disabilities around the country. In an open letter to the members of the Foundation for Science and Disability, the deaf microbiologist John J. Gavin (1979a), then president, argued against the Supreme Court perspective:

> Thus, they suggest institutions of higher learning have the *right* and the *ability* to determine whether a handicapped person can qualify for a profession on the basis of their physical status regardless of their academic qualifications. We take issue with this reasoning because scientific and technical work can be exclusively intellectural in nature, scope and content. The *sine qua non* of such work is intellectual competence, and one might logically assume that substance would take precedence over the form in which it is packaged. (p. 4)

Gavin was deafened in 1956 as a result of an adverse reaction to dihidrostreptomicin at the age of thirty-three. He was persuaded of the need for activism when he himself experienced discrimination at ten universities to which he had applied for Ph.D. work. In addressing the Davis decision, Gavin explained how nursing is no different from many other professions which require a license for practice. While the licenses are general, the practice is often limited; and specialization[9] is common, with individuals never expected to perform every task for which they are licensed:

> Nursing is no exception. There are any number of specialties a nurse can select when hearing problems would pose little or no difficulties. But to work in a specific area requires certification as a registered nurse. And to

> qualify as a registered nurse, certain formal academic course work must be completed. To deny an individual the opportunity to acquire the fundamental knowledge upon which the practice of any profession is based because at some future time there might be a danger to public and individual safety because of physical disability is unreasonable. (p. 5)

The Supreme Court ruling even had an effect on the lives of deaf nurses already practicing. Deaf since childhood, Keith Ragsdale graduated from a nursing school in Corpus Christi, Texas, and has worked at several hospitals in Texas and Indiana since 1971. Ragsdale traveled the country in the wake of the Supreme Court decision to show that deaf individuals, as well as those with other sensory and physical impairments, *could* become nurses (*New Haven Register*, November 18, 1979). Yet, available data suggest that the situation in nursing has not improved much since the Davis decision, and, as in the case of deaf physicians, deaf nurses face what seem insurmountable obstacles in both their training and in employment.

These legal difficulties and attitudinal barriers are also encountered by deaf Europeans. In May, 1991, I attended a conference in Heidelberg, Germany, where a medical doctor, deaf since the age of six, presented a paper entitled "University—And Then?" Roland Zeh, born in Esslingen in 1960, completed his medical studies at the University of Freiburg in 1989, and in 1991 he was a doctor in practical training in Worms, Germany. In discussing the legal difficulties he faced, he mentioned that a license cannot be given if an applicant is "due to a mental or physical disability, not able or suited to pursue the medical profession." Zeh explained that "a very narrow interpretation of such a passage can result in serious difficulties" for the aspiring deaf medical student:

> I have been, for example, denied [admittance] to the final examination of my medical studies because the authority had become set on the view that a doctor has to be able to fulfill all the tasks of a doctor, including, for example, auscultation. The fact that there are also many medical occupations which do not require auscultation, was not disputed by the authority. It is true that I was allowed to take my examination after legal proceedings in the second instance but the problem is not resolved yet since the paragraph [legal reference to "physical disability"] is still in existence and since there has been no leading decision yet. (Schulte & Cremer, 1991, p. 7)

Zeh expected to face many legal obstacles in the future, and that the barrier for deaf people in Germany is similar in other professions. "The dilemma," he said, "is that we hearing-impaired people only have a chance if we are better than others in a special field. In order to reach this goal, however, we often have to remove irreconcilable obstacles which are of no use to the cause" (p. 8).

Removing such obstacles has been the primary purpose of several professional organizations established in the United States. In 1966 the American Professional Society of the Deaf (APSD) was founded. Among its objectives was the provision of role models for inspiring young deaf men and women toward professional careers and working toward the eradication of persistent misconceptions about deafness and deaf people. During the early years the society members met at the Engineers' Club in New York City, then later at the Institute of Rehabilitative Medicine of the New York University Medical Center. They served as an advisory group for deaf people interested in professional careers. APSD was actively involved with the American Association for the Advancement of Science (AAAS), helping to provide in 1975 a symposium titled "The Physically Disabled Scientist: Potential and Problems." Members of the APSD included the chemist Edgar Bloom, the engineer Albert Hlibok, and Donald L. Ballantyne, professor of microsurgery. Although the society lost momentum after a decade, it had a great impact when it was much needed, particularly in relation to its "deaf awareness" work.

The Foundation for Science and Disability (FSD) has also served an important role in enhancing the professional lives of scientists. FSD is affiliated with the AAAS and holds its annual meeting during the AAAS convention. Pioneering leaders included hearing scientists with disabilities such as Ed C. Keller, Jr., and S. Phyllis Stearner, and the deaf microbiologist John J. Gavin, director of allergy research affairs for Miles Laboratories in Elkhart, Indiana. Over the years, FSD has provided support for scientists with disabilities in seeking scientific positions. The organization has also provided financial support through awards to students with disabilities interested in science careers. FSD shares information with its membership on access and assistive device issues, and was very active in 1993 in assisting the National Research Council with the development of science education standards for curriculum, teaching, and assessment.

Whether the participation of deaf men and women in science will increase as a result of such efforts depends in large part on how the academic preparation in elementary and secondary programs improves over the next few decades, and whether deaf people continue to challenge discriminatory attitudes and practices in educational and occupational settings. This political voice of deaf persons, and continued support from hearing people, are of paramount importance.

Although the prevailing leverage of the deaf professional has, by and large, remained weak, several phenomena of the modern period have marked a reformation in society and a promise of better days ahead. One has been the progress in technology, which has enhanced communication (the telephone coupler, for example) and access to information (the television captioning decoder). The second was the "civil rights" movement for people with disabilities, which led to federal legislation and further access to employment and education opportunities. And the third was the 1988 protest at Gallaudet University, indicative of increased political voice among the Deaf community. Each of these has helped significantly to upgrade the leverage of deaf men and women professionals, including those in science, and continues to have positive effects on patterns of employment.

TECHNOLOGICAL CHANGE AND THE ISSUE OF ACCESS

In 1963, when I was profoundly deafened, the loss of hearing seemed very "disabling" to a young teenager, not only in the sense that so much adjustment would be necessary in emotional and psychological terms, but also in terms of my thinking about a career. Communication was, of course, paramount in my thoughts. Gone was the telephone, which I had taken for granted. It would be several more years before Bob Weitbrecht would market the acoustic telephone coupler for use with a teletypewriter (TTY). The Deaf community had yet to experience the struggles in obtaining the cumbersome machines and to convince the public of the potential of a network of TTY users. This period in the "deaf experience" in science and technology was analogous to the development and implementation of telegraphy a century earlier, particularly in regard to public attitudes and acceptance.

In the same sense, I remember watching television for twenty years with little or no understanding of what was being said. I sat through televised reports on the assassinations of the Kennedy brothers and Martin Luther King, Jr., the Cuban Missile Crisis, and the Apollo lunar landing . . . in complete silence. I recall writing to one network, inquiring whether officials might consider showing subtitled versions of foreign films on television during the early morning hours, rather than the dubbed voice versions, only to receive the reply that the subtitles were found to be distracting to hearing viewers. It would be years until the development of the captioned television decoder would bring more awareness and information into my world. As a young boy I was also an avid fan of shortwave radio, sitting in my bedroom at night while listening to the hams and other stations around the globe. Suddenly, I found my equipment to be a meaningless collection of electrical circuitry. Years after I became deaf, radio TTY grew in popularity and, today, a deaf enthusiast can use computer software to convert Morse code to alphanumeric characters on the screen and, in a sense, "listen" to one form of radio communication.

The profound impact inventions have had on the lives of deaf people over the past few decades is very much a part of the "deaf experience" in the history of science, precipitating concomitant social change in the modern period. Both boon and nemesis, however, technology once again had a discouraging effect on the Deaf community in the 1960s, this time in the form of automation. About 60 percent of deaf male workers at this time were craftsmen or machine operators, and the impact of automation was especially felt (Tully & Vernon, 1965). When computer technology began to show promise, some deaf people entered this field, but by 1970 there were only about 150 deaf people employed as programmers, approximately half of them in the Washington, D.C., area. Acceptance in the computer field was much too slow for deaf people and, as Robert L. Bates (1970) summarized, the "difficulties and frustrations encountered by these potential programmers qualify, quite frankly, as discrimination" (p. 1). When the space age began, additional opportunities for deaf people opened up, albeit still few. Lockheed Missiles and Space Division of Lockheed Aircraft Corporation in Sunnyvale, California, employed about forty deaf technicians and even developed a manual of technical sign language in the early 1960s for supervisors and deaf employees.[10] Victor H. Galloway was a systems engineer for twenty years at Lockheed. Deaf technicians were also hired for the Polaris project, at Philco's Palo Alto plant to assemble the Courier I-B satellites, and in other space projects. T. Alan Hurwitz was an associate electronics engineer in the Plasma Physics Department of the Aerospace and Aeronautics Center at McDonnell Company in St. Louis, Missouri. He conducted computer analyses of communications and radar systems performance of space capsules using the ray-tracing techniques in the Gemini project, and he was responsible for investigating methods of directing ground antenna to recapture the refracted radio waves of a space capsule entering the ionosphere from outer space. During his years at the Stanford Research Institute (1958–1969), the physicist Robert H. Weitbrecht conducted studies on high-definition photography of the Echo satellite and resulting work on atmospheric scintillation. Another deaf physicist, Robert S. Menchel, worked on a project studying the heat shields for the Apollo capsules. Menchel also helped to develop instrumentation for geophysics and astrophysics, including a control monitor which activated one of six transmitters upon request from a master control station. Robert L. Bates was responsible for U.S. Navy projects involving ballistic entry for satellites and rocket instability.[11]

The Deaf community in the United States watched curiously as a group of deaf men assisted the United States Naval School of Aviation in Pensacola, Florida, with its tests on the effects of prolonged exposure to space flight conditions. Concerned that no one should be sent into space handicapped by functional symptoms resulting from an environment of prolonged exposure to weightlessness or to constant rotation, Captain Ashton Graybiel's staff from the U. S. Naval School of Aviation Medicine had visited Gallaudet College for the Deaf in Washington, D.C., in search of subjects for their investigations. Their concern was justifiable. Some of the personnel involved with manned space efforts in both the United States and the Soviet Union had been confronted with inner ear problems and motion sickness. In 1963, Alan B. Shepard, America's first person in space, was grounded with Ménière's disorder after just one space mission. The problem was corrected surgically by Dr. William F. House, director of research for the Los Angeles Foundation of Otology, who placed a small plastic tube in Shepard's inner ear to equalize the pressure of the fluids. Shepard regained his space pilot status and continued to fly. Graybiel's research staff sought deaf volunteers with no functional labyrinthine canals who would be willing to communicate their feelings and sensations for the medical staff during extensive experimentation. The deaf subjects spent several days of continued whirling in a centrifuge. Those around them with normal hearing dropped out with motion sickness, but the deaf participants came out of it showing no effects except unsteadiness in their legs. They underwent tests in a tilt box in complete darkness, and they sat through experiments in another machine which rotated both vertically and horizontally at the same time. In addition, they spent one night riding an express elevator up and down the Empire State Building in New York, experienced weightlessness in aerial maneuvers, and rode through a violent storm in a small boat off the coast of Nova Scotia. In the latter experiment, they battled the heaving sea, forty-knot wind, and extremely low temperatures for twenty-eight hours. The deaf men suffered only from fear and imbalance, while most of the hearing participants on the boat became ill.

Access to the Telephone: The Acoustic Coupler

The breakthrough that enabled deaf people to use the telephone in a practical way came in the 1960s, and three deaf men—a physicist, a dentist, and an engineer—were responsible for patenting and marketing the ingenious device. That "necessity is the mother of invention" was never more true than in the story of the development of the acoustic coupler for the telephone.

In 1963, Robert H. Weitbrecht was climbing Mount Lassen with a friend when he paused near the top of the volcano to take in the beauty of the mountain view. There, Weitbrecht encountered the McKeowns, a couple who were also hiking, and who had heard Weitbrecht's voice nearby. Having a deaf son of their own, the McKeowns were familiar with the "distinct voice" of a deaf person (*DA*, *32*[10], 1980) and introduced themselves. Upon returning home, the McKeowns immediately contacted Arthur B. Simon, a friend of theirs, who subsequently communicated Weitbrecht's interest in telegraphing to James C. Marsters, a congenitally deaf orthodontist from Pasadena. In a letter Marsters wrote to Weitbrecht on April 26, 1964, he planted the idea of using the telephone in combination with teletype:

> What I have in mind, Bob, is the possibility of a network of regular telephone line RTTY for deaf people who can afford one . . . but not to lease one via the telephone company nor a special telephone line. You do have a special telephone to Larry, don't you? If so, why won't it be possible to translate over the regular line by proper modification of equipment . . . granted that eventually there will be enough units? Of course, regular RTTY would be better for us, but I think of the many deaf people who do not have the time, money, etc. for a general license. What do you think?

Marsters then flew to San Francisco, and, under his encouragement, Weitbrecht began to experiment extensively. Both Marsters and Andrew Saks, a deaf engineer with experience in the business world, provided Weitbrecht with suggestions for his design of the device. For their investigations, Marsters purchased among other things a new Model 32 ASR teletype machine. They built a successful "hardwire acoustic converter" which would produce for each key on the teletypewriter an audio signal for transmission over the telephone. A similar transmitter-receiver unit on the other end of the line would decode the tones and the message was then printed on another teletypewriter. The device did not render the telephone system inoperative for ordinary use, since the phone was placed on a "coupler" (cradle) containing a loudspeaker and an induction coil. This invention in 1964 allowed deaf people to finally use the telephone after nearly a century of frustration.

Weitbrecht, Marsters, and Saks then began a partnership to spread the use of teletypewriters among deaf people. They had no government funds. In the 1960s, the U. S. Department of Health, Education, and Welfare was interested in voice-to-print research, and encouraged the three men to pursue such work. But the need for use of the telephone by the grassroots Deaf community was a priority to them. Weitbrecht resigned from the Stanford Research Institute in 1966 to work full-time on the development of the coupler. In 1967, the Applied Communications Corporation was formed to market the coupler, with Saks as president. The three deaf men invested their own money in the research and development, with no government or private financing.

Once the device known as the Phonetype was ready to distribute, the Deaf community responded. All over the country, deaf people volunteered for retailing, installation, and servicing teams. At the American Telephone and Telegraph Company corporate secretary's office in 1967, Weitbrecht, Marsters, and Saks demonstrated the system and suggested that the Alexander Graham Bell Association for the Deaf (AGBAD), the National Association of the Deaf (NAD), and other organizations of the Deaf community be the recipients of old teletypewriter equipment. AT&T and the Bell Association then assisted in the search and delivery of the bulky teletypewriters. In Indiana, the Teletypewriters for the Deaf Distribution Committee, established in 1968, changed its name to Teletypewriters for the Deaf, Incorporated (TDI), and continued to locate more machines for this purpose and to organize a national directory of users. Jess Smith from NAD and H. Latham Breunig from AGBAD were leaders in these efforts. Using computers, they distributed employment opportunities, emergency phone numbers, and news for deaf people over the phone lines. The Western Union Telegraph Company and other organizations donated thousands of used teletypewriters to the Deaf community. Paul Taylor, a deaf chemical engineer and close friend of Weitbrecht's, was instrumental in establishing a teletypewriter network in St. Louis, Missouri, and the Midwest, for example. Taylor and his wife, Sally, wrote a maintenance manual for the machines. Irvin Lee Brody, unable to enter medical school after completing premedical work at Rutgers University, decided to dedicate his time and energy to help establish the New York-New Jersey telephone-TTY network, at the time the largest in the nation. Within a decade, there were two hundred local agents (for installation and repairs) and over 15,000 deaf telephone-teletypewriter users.

The equipment being used was workable with either ASCII or Baudot transmission codes, but due to a pending legal case, AT&T suggested that the Deaf community avoid use of ASCII frequencies.[12] ASCII equipment and technology was also more expensive and complicated in the 1960s. Baudot was not compatible with ASCII machines. Advances in electronics soon led to a variety of compact models to which Weitbrecht and his partners contributed dual ASCII/Baudot compatibility with LED displays or paper printouts. More recent telephone devices include voice synthesis chips and compatibility with personal computers, making the machines even more versatile.

In 1966, Marsters was invited to demonstrate the TTY system at the University of London at his own expense, but by 1972, the Deaf community in England was still unable to convince the British Post Office, which controlled the telephone network, to permit the use of the acoustic couplers. Weitbrecht and Saks then traveled to England and were successful in gaining approval. Through the efforts of these three deaf men, use of the TTY soon spread to Switzerland, New Zealand, Australia, and other countries, and TDI (which changed its name to Telecommunications for the Deaf, Incorporated), eventually experienced international success.

Yet even with these improvements many difficulties were experienced with the telephone. Unless both the caller and receiver had TTYs or one of the electronic devices then coming onto the market, direct phone conversations were not possible, and many deaf people were forced to continue with third party interpreting. Experiments with picture telephones proved such devices not feasible in the early 1970s because of the requirements for special cables to carry the video signal. Various companies also experimented with telewriters, code-sending devices with visual displays, and vibrotactile apparatus, but again there were requirements for compatible equipment at both ends of the line. In the 1990s the establishment of telephone relay services rapidly expanded the accessibility of the telephone, allowing deaf people to type their messages while the relay service personnel spoke to the party having no special phone device. Similarly, anyone could call a deaf person by voice by dialing the relay service, which subsequently typed the message.

Access to Television: The Closed-Captioning Decoder

The development and implementation of closed-captioning technology similarly required persistence on the part of the Deaf community in the face of attitudinal barriers. In 1947, Emerson Romero, a deaf actor who had left Hollywood with the advent of the talkies, began to splice homemade subtitles to short films he had purchased. The promise of access to captioned films grew. In 1958, President Eisenhower signed Public Law 85-905 creating the Captioned Films for the Deaf Program, with Malcolm J. Norwood as the chief of Media Services and Captioned Films for the Deaf Department of the Bureau of Education of the Handicapped, the first deaf person to hold a professional position in the Department of Education. Over the next decade, Norwood investigated the potential of captioned television, meeting resistance from hearing people over the use of open captions. Engineers at the Caption Center at WGBH in Boston were then awarded a contract to develop the first caption computer. In 1971, the first "closed-captioning" demonstration took place in Knoxville, Tennessee, and the U.S. Office of Education supported the development of the captioning decoder. Through the 1970s, the use of closed captioning grew slowly, with special cooperation from the ABC television network and Washington, D.C.'s PBS station, WETA. The Federal Communications Commission agreed to reserve line 21 of the television signal for captions in 1976, and NBC and CBS soon joined in. The National Captioning Institute (NCI) was established in 1979 as a private, nonprofit corporation to expand opportunities for deaf people through television access. With 600 retail outlets carrying the decoders by 1988, captioning technology was finally able to have a positive and significant impact on the lives of deaf people.

The Television Decoder Circuitry Act, signed into law October 16, 1990, required closed-caption decoder circuitry to be installed in all new television sets with thirteen-inch screens or larger sold in the United States by July 1, 1993. The "Chip Bill" guarantees access to a powerful information medium for the more than 24 million deaf and hard-of-hearing people in the United States.

While the past 100 years have been marked by exciting breakthroughs in assistive devices, and in telephone, television, and computer technologies, deaf people have learned to be ever cautious. As has been shown in this book, barriers were created for deaf people, rather than broken down, with the invention of the telephone, the development of radio, the advent of "talkies," and the impact of automation. Remaining proactive as new technologies develop over the next decades will help avoid such barriers. Voice mail and other voice-based technologies, for example, may present an obstacle. Deaf people should work collaboratively with these researchers to assure that new products and technologies do not discriminate against people with speech impairments. And, finally, experience with federally funded research on captioning technologies has shown how dreams can become reality. A stronger thrust toward federally funded research in speech-to-text technologies will bring deaf people all the more close to having careers on an equal basis with hearing colleagues.

FEDERAL LEGISLATION AND THE REMOVAL OF BARRIERS

In the face of continued prejudice and discrimination, much more than technological breakthroughs was needed to change attitudes in the years after World War II ended. The 1960s and 1970s were particularly exciting years for the Deaf community in America, revolutionary in a sense, with the impetus of the civil rights movement and federal legislation in the vocational rehabilitation and education arenas. Additionally, increasing respect and recognition were being given American Sign Language (ASL) in school environments, and this was accompa- nied by a growing political voice among Deaf people. Along with the technological advances, the social and political transformations which took place have led to wholly new lifestyles for many deaf people as well as improved attitudes about deafness in general.

Vocational Rehabilitation

Inspired by the great gains experienced as a result of the Civil Rights Act of 1964, people with disabilities pressed for their legal rights, and changes in vocational rehabilitation and other legislation soon followed. The Vocational Rehabilitation Act of 1954 had provided much-needed federal support for Americans with disabilities, particularly in allowing states to develop rehabilitation centers. Among the Deaf community leaders actively involved in implementing the recommendations was Boyce R. Williams of the Deafness and Communication Disorders Office of the Department of Health, Education, and Welfare. With Mary E. Switzer's support, Williams provided funding assistance to start such organizations as the Registry of Interpreters of the Deaf, the Professional Rehabilitation Workers with the Adult Deaf (now the American Deafness and Rehabilitation Association), and the Council of Organizations Serving the Deaf. The act also led to funding for the Leadership Training Program in the Area of the Deaf at San Fernando Valley State College in Northridge, California. Subsequent amendments extended federal support. One important piece of legislation was the Architectural Barriers Act, passed by Congress in 1968, which directed federal agencies to establish minimum standards in regard to visual warning signals in the workplace, and amplification systems in telephone and meeting areas. Changes were too slow for many individuals with disabilities, however, and in the Washington, D.C., office and regional offices of the Department of Health, Education, and Welfare, a sit-down strike was necessary to pressure for implementing the regulations of the Rehabilitation Act of 1973. Under Section 502 of this act the Architectural and

Transportation Barriers Compliance Board was established, and among its responsibilities was the provision of technical assistance to groups affected by regulations. For deaf people, issues of physical accessibility focused primarily on the provision of telecommunication devices. Section 504, which prohibits discrimination against qualified people with disabilities on the "basis of handicap," was a major turning point in the legal rights movement for persons with disabilities in the United States.

Education

In 1975 Public Law 94-142, the Education of All Handicapped Children Act, was signed by President Ford and guaranteed free, appropriate public education for all children with disabilities in the country. In 1977 the White House Conference on Handicapped Individuals was held to generate a list of recommendations for additional improvement in the quality of living and working conditions for people with disabilities. PL94-142 has been a catalyst for educational research, curriculum development, and the promotion of active involvement and change in professional organizations concerned with science and science education. Probably in no other discipline has such a rapid and successful reaction to PL 94-142 swept through the ranks of educators and other professionals as in science. Prior to PL94-142, in 1973, the Board of Directors of the American Association for the Advancement of Science (AAAS) established the Office of Opportunities in Science. The Project on the Handicapped in Science was launched in 1975 and has resulted in a voluminous amount of information being published and disseminated on the topic of science education for students with disabilities. Since that time, the AAAS has held many national conventions with regularly offered sessions pertaining to science as a career for people with disabilities. Other professional organizations such as the Association for the Education of Teachers in Science (AETS), the American Chemical Society (ACS), and the American Association of Physics Teachers (AAPT) have also offered special sessions, colloquia, or publications on the teaching of science to students with disabilities. The American Chemical Society devoted a major part of an issue of *The Journal of Chemical Education* to ten articles on teaching chemistry to students with disabilities. The AAPT Symposium on the Disabled in Physics in 1979 included collaborative presentations by physics professors from Gallaudet University and the National Technical Institute for the Deaf (NTID). In 1981, the AETS held its convention at NTID. In addition, the National Education Associaton (NEA) published a textbook entitled *Teaching Handicapped Students Science* in 1981. The effects of PL 94-142 and the efforts of these organizations are also apparent in the increase in published articles after 1976 on teaching deaf students in science.

New organizations have also been established to facilitate participation in science of students with disabilities. The Science Association for Persons with Disabilities (SAPD), formerly the Science for the Handicapped Association, is affiliated with the National Science Teachers Association. Dr. Ben Thompson was a founder and an influential leader in this organization for many years.

SAPD has provided much information and consultation to science educators in many countries.

The National Technical Institute for the Deaf

Public Law 88-565, the Vocational Rehabilitation Act passed by Congress in 1954, provided many services to people with disabilities. For deaf people, this included funding for continuing education in university or technical training programs. Yet, with all of the exciting programs and changes occurring in the decade to follow, opportunities for technical education remained largely inadequate. Public Law 89-36, signed by President Lyndon B. Johnson in 1965, addressed this great need through the establishment of the National Technical Institute for the Deaf (NTID) at the Rochester Institute of Technology (RIT). NTID admitted its first students in the fall of 1968. Robert Frisina was the institute's first director and William E. Castle the first dean. NTID now enrolls more than 1,100 deaf college students in various scientific and technical career programs. As one of the nine colleges of RIT, NTID provides direct instruction in many technical majors, as well as support services for deaf students enrolled in programs in the other eight colleges. Deaf students interested in science or engineering careers, for example, cross-register in the RIT Colleges of Engineering, Science, or Applied Science and Technologies, earning college degrees in engineering, science, and mathematics-related fields of study. Graduate follow-up data indicate that of all the deaf RIT graduates who enter the work-force, about three out of ten hold positions as scientists, engineers, computer scientists, or statisticians, or work in clinical, electrical, and other scientific laboratories as technologists and technicians. Among the alumni of NTID are Andrew Baker, who earned a doctor of optometry degree after completing his bachelor's degree in biology through NTID at RIT, and Alan C. Gifford, a senior field cost engineer in Brockton, Massachusetts. Other graduates have become secondary and postsecondary teachers in science and mathematics programs. One of these graduates, Fred Mangrubang, won the 1990 Presidential Award for Excellence in Science and Mathematics Teaching sponsored by the National Science Foundation, the only deaf person to have received this honor.

Along with NTID, regional programs for deaf students have been established at the Technical Vocational Institute in St. Paul, Minnesota, the Washington Community College in Seattle, Washington, and the Delgado Vocational Technical Junior College in New Orleans, Louisiana. The California State University at Northridge also began a program, including a National Leadership Training Program in the Area of Deafness. R. L. Hoag (1989) estimated that by the end of the 1980s there would be about 11,000 deaf students in colleges and universities in the United States. Of these, 1,800 would be Gallaudet students and 1,200 NTID students, with the other 8,000 enrolled in programs throughout the country, including "non-degree vocational and technical programs, two- and four-year technical and liberal arts programs, and graduate programs" (p. 94).

The Americans with Disabilities Act

The Americans with Disabilities Act (ADA) of 1990 has been the most important civil rights legislation for people with disabilities in the history of the United States. Much stronger than earlier legislation, the ADA provides employees with disabilities with legal protection against discrimination in public and private areas. For deaf people, this means greater access through interpreters and telecommunication devices. Public service announcements produced by the federal government must be closed-captioned. Through the ADA the responsibility for "reasonable accommodation" is clearly on the shoulders of employers and service providers. Known as the "Emancipation Proclamation of the Disabled," the ADA provides more than 43 million people with disabilities the same civil rights protection that previous legislation provided on the basis of sex, race, religion, and national origin. Access to mass transportation, public accommodation services, and state and local government are key issues in the act. To respond to the growing demand for information about the ADA, the President's Committee on Employment of People with Disabilities expanded the Job Accommodation Network established in 1984. Building upon the substantive framework of Section 504 of the Rehabilitation Act of 1973, the ADA renewed attention to issues of access and programs on college campuses as well. This will hopefully increase the number of qualified students with disabilities in colleges and universities and possibly help in alleviating anticipated shortages in the job market, including scientists and engineers.

Title IV of the ADA mandates accessibility to communication by telephone. Telephone companies are required to provide dual party telephone relay services for local and long-distance calls, thus enabling deaf and speech-impaired persons to have equal access to telephone services for the first time in history. Emergency 911 systems must be accessible with direct TTY lines. Paul Taylor, a deaf scientist, testified before the U.S. Senate in support of the bill. "Title IV will have the greatest impact in the employment field," Taylor explained, "since the telephone is probably the biggest impediment to job mobility and career opportunities for hearing-impaired people" (*Silent News*, *22*[8], 1990, 4).

The inroads made by these and other legislative acts of the 1960s and 1970s led to a much more encompassing and politically sophisticated social movement that has significantly affected the lives of deaf people in the United States.

NEW POLITICAL VOICE: THE GALLAUDET UNIVERSITY PROTEST OF 1988

When deaf students began demanding the first deaf president at Gallaudet University in early March, 1988, the news spread quickly around the world. I was at this time a visiting lecturer at the University of Leeds, England. In the evenings, I conducted some research for this book. It was purely by coincidence that at the very moment news broke out about the protest, I was studying the life of Frederick Augustus Porter Barnard. How ironic it seemed to me that Columbia University (then Columbia College) had flourished for many years under a scientist who was functionally deaf, a skilled signer, and a former teacher of deaf pupils, while Gallaudet University had never had a deaf president. Granted, Barnard tried fruitlessly to communicate, without the help of a sign language interpreter, through the use of a large mechanical amplifier which he had contrived. Among his great accomplishments was his vigilant appeal for the education of women, and he is well honored for it in the name of Barnard College, incorporated six months after his death in 1889. Now, a century later, Gallaudet students were organizing the "Deaf President Now" protest, barricading the entrances and emotionally calling for resignation of Elisabeth Zinser, a hearing scholar and administrator, but a person with no experience in the Deaf community who had been selected over two capable deaf finalists. I was frustrated that I could not go with my students from NTID to join the protest, which closed down Gallaudet University for a week. The demands that Zinser resign and that 51 percent of the board at Gallaudet University be made up of deaf persons were met. I. King Jordan became the first deaf president of Gallaudet University.[13] The message behind this protest soon became clear. Ripples of reform have since been observed across the country, with more and more qualified deaf professionals being appointed to leadership positions. The empowerment of deaf people in their own education has dramatically gained a formidable foothold as a result of this experience.

Conclusion

THE ENIGMA OF ATTITUDES

In June of 1978, I attended a working conference entitled "Barriers to Postsecondary Science Education for Handicapped Students" sponsored by the American Association for the Advancement of Science (AAAS) Office of Opportunities in Science. The goal was to identify barriers which inhibit students with sensory and physical disabilities in their pursuit of careers in science and to propose solutions to breaking down these barriers. Also attending this conference were other scientists and university students with disabilities, counselors, university administrators, and science professors. There were six categories of barriers identified: informational, financial, attitudinal, environmental, communicational, and academic. Informational barriers, for example, include the lack of resources and methods, or knowledge about them, which makes it difficult for a student with a disability to receive information. As an illustration, blind students may not have access to scientific writings in Braille, and educators may not be aware of alternative methods for providing this information to them. Similarly, a deaf college student may be challenged by a learning experience based on the use of audiotapes. An informational barrier is created if the audiotapes are the *only* available form in which the student may learn, as well as if there are transcriptions available not known to the professor or student. Environmental barriers would include, for the deaf student, the failure to provide a warning light indicating a fire alarm has sounded. Communication barriers for deaf people are many, as has been discussed in this book, and there are other communication barriers for people with impaired sight or speech, even those with normal hearing. Academic barriers, as summarized by the AAAS (Redden, Davis, & Brown, 1978), involve challenges and problems which necessitate adapting teaching and evaluation strategies and instruments, as well as the amount of time permitted to complete tests, projects, and other course work. Financial barriers relate to the extra expenses incurred in overcoming other barriers.

In a sense, however, all of these categories have their roots in the enigma of attitudes, and there was strong agreement among participants in the AAAS conference that "attitudinal barriers" were the most serious of all the types. Many kinds of attitudinal barriers were identified, including lowered expectations, categorical thinking (one deaf student's failing a physics test may be interpreted as implying that all deaf students have difficulty in physics), reluctance to ask and answer questions, attitudes about accommodation to a disability, assumptions of inferiority, assumptions of responsibility (the assumption that persons without disabilities are responsible for the safety of students with disabilities), spread of effect (a person who is deaf is considered unable to think, for example; and the anachronism "deaf and dumb" is still encountered today in various forms), and stereotyping (e.g., deaf people can read lips around corners, or they have a phobia about science and mathematics). As one deaf scientist reported at the conference, "I was told deaf people have trouble learning mathematics, that there was no need for me to take it in high school" (Redden, Davis, & Brown, 1978, p. 11).[1]

As has been discussed in this book, a loss of hearing is not so much an obstacle to most deaf people as are the *attitudes about deafness* which they face. In this sense, the meaning of "handicap" as a "hindrance" comes into play. Unfortunately, this enigma of attitudes includes preconceived, unjustifiable notions among educators and scientists about people who are deaf. One that I have frequently encountered is the belief that people with congenital and early-onset deafness are less able to succeed in scientific careers. It is an attitude that reflects not only our failure to adequately educate our students, but also indicates how little we know about the achievements of deaf professionals with congenital and early-onset deafness. In effect, age of onset of deafness and amount of hearing loss are factors which become stigmatizing at times for people who are deaf, but they are not traits which should preclude science as a career for deaf people. As this book reveals, certain personal qualities important to success, including fortitude, perseverance, self-advocacy, and strategies for dealing with accommodation and accessibility, are attributes needed in defense against such negative attitudes.

On the basis of the sample of 675 deaf scientists identified while writing this book, the data indicate that the distribution of deaf men and women scientists has been slowly broadening from a primary interest in chemistry to a more widespread distribution in biology, engineering, mathematics, and medicine. Doors are slowly opening to deaf persons in nearly every field of science and medicine, although, in comparison with hearing persons, the percent of deaf people entering all fields of science continues to be extremely small.

RECOMMENDATIONS

For deaf people to gain meaningful access to scientific careers, a number of issues must be addressed over the next decade. Seven of these issues described below are (1) curriculum development in science, (2) teacher preparation, (3) occupational stereotyping, (4) attitudes of "significant others," (5) job satisfaction, (6) access to professional organizations, and (7) the need for continued activism.

Curriculum Development in Science

Despite the fact that teachers and administrators have been working hard to improve curricula and teaching, science and mathematics remain low-priority subject areas for deaf students today. Curriculum research and development is strongly needed through collaborative efforts of universities, local school systems, and scientific organizations. Approximately 30 percent of deaf students attending high school exit without a certificate, diploma, or other credential (U. S. Department of Education, 1990). Of those who complete high school, between 10 and 20 percent receive certificates which do not qualify them for postsecondary education, and seven of every ten deaf students who are admitted to two-year and four-year colleges and universities withdraw before earning a degree. Reasons for the retention problem range from inadequate academic preparation to unhappiness with the social environment.

Teacher Preparation in Science

A majority of the science teachers deaf children have encountered in elementary and secondary programs are not adequately prepared in science. Most of these science teachers have degrees in the field of educating deaf students, but they often do not know science (Lang & Propp, 1982). In Corbett and Jensema's (1981) study of teachers in school programs serving deaf children, they identified 2015 teachers responsible for science classes out of 4887 respondents, but only 165 (3.4 percent) of these 2015 teachers had undergraduate majors in the physical sciences. In addition, only 3.7 percent of the males and 0.6 percent of the females reporting state certification were certified in teaching science.

Despite this problem with teacher preparation, deaf students have positive attitudes toward science. Lang and Meath-Lang (1985) administered a five-item questionnaire to 329 deaf junior and senior high school students in nine school programs in order to identify reasons why these students liked or disliked science. Nearly 86 percent of the students in this survey responded that they liked science. Reasons for liking science included specific topics in the courses (48.7 percent), instructional strategies with particular emphasis on "hands-on" experiences (19.2 percent), and the recognition of the importance of science in their future (10.8 percent). Only 2 percent of the responses to the question specifically targeted the teacher as a reason for liking the subject. However, when asked directly if they felt their science teachers were good, 80 percent of the students responded affirmatively. The ability to clearly explain concepts, effectively communicate in sign language, and use a variety of strategies in teaching science were the most commonly mentioned characteristics of good teaching in science.

Occupational Stereotyping

Deaf students seldom have the chance to meet successful deaf scientists, or even deaf science teachers, as models. About one in ten teachers of deaf students in K-12 programs in the United States is deaf (the ratio is much lower in other countries). Few of these deaf teachers are employed in public mainstream programs where many deaf children are integrated with hearing children. Without models to emulate, deaf students form their own career-related notions about which professions are "appropriate" for them. In 1977–1978, Robert S. Menchel travelled throughout the United States as a deaf model scientist in a project sponsored by the American Association for the Advancement of Science and Xerox Corporation. Meeting with over 3,500 deaf children, their parents, and teachers, Menchel described his experience as a physicist and encouraged the children to think about science as a career.[2] Some young deaf children who grow up never having the opportunity to meet deaf adults develop the belief that they will be hearing when they become older; others believe they will not reach adulthood. In one of the many letters written to Menchel which he shared with me, a young deaf girl expressed surprise about meeting him: "I couldn't believe that so many deaf people has [sic] the courage to have those important and good jobs. . . . I never realize that a person like you or other deaf people can really get great jobs." For children such as her, there are few deaf women scientists and science teachers available, and this may be a factor influencing how young deaf women perceive science as a possible career.

Ironically, occupational stereotyping by at least one international leader in the education of deaf students at the beginning of the twentieth century actually *favored* science as a profession for deaf people. In 1905, Giulio Ferreri published a report comparing higher education in Italy and the United States and their potential for educating deaf youth. "I believe that as the Blind must content himself, when he places himself on an equality with normal persons in the higher schools," wrote Ferreri, "to follow the courses in literature, history, and philosophy, just so the Deaf must limit his aspirations to the study of the branches of physics and mathematics in their mechanical and esthetic application" (*VR*, *7*[1], 1905, p. 27). Perhaps, based on purely logistical considerations, there is an element of logic in this quaint generalization. I recall my own experiences as a physics major, and the reflections of my deaf colleagues who also studied in college in the days before sign language interpreters were available; some of them *had* chosen engineering, mathematics, or science majors because of the visual nature of the work.

Attitudes of "Significant Others"

A fourth target for change should be the attitudes of those responsible for making decisions about admitting deaf students into college programs, or accepting deaf professionals into employment positions. Those responsible for advising deaf students about possible careers have also demonstrated attitudes which present barriers to the pursuit of science. The documented literature on the resistance deaf people have encountered in entering fields of science, or in matriculating into colleges and universities, is largely narrative in nature, but is abundant enough to substantiate this issue as problematic and in need of attention. The attitude that a science major for deaf students is inappropriate has even led some college admissions personnel and vocational rehabilitation agencies to reject applications for financial assistance by deaf students (Redden, Davis, & Brown, 1978, p. 19).

In his presentation to the AAAS, Jerome D. Schein (1975) provided an example of employer discrimination. Schein, director of the Deafness Research and Training Center at New York University, recalled that a distinguished deaf scientist with an international reputation told him that he was not permitted by his employer to drive a company car. Even after Schein provided data on the superior safety records of deaf drivers, the employer's attitude did not change.

Job Satisfaction

Occupational factors such as underemployment, discriminatory experiences, difficulties in communicating on the job, and a lack of opportunities for upward mobility have discouraged some deaf scientists from remaining in the profession. In my interviews with deaf scientists and engineers who left scientific positions to enter educational environments, the lack of opportunity for upward mobility, in particular, was a frequent explanation.

Access to Professional Organizations

Deaf scientists face difficulty in gaining full access in professional science organizations. As the molecular biologist Jane Dillehay has explained, a deaf scientist has a greater challenge in keeping up with research: "I can't take part in scientific meetings the way I want to. First, it is hard to be accepted at the meeting. Second, no matter how qualified, the interpreters don't understand the technical aspects of the topic, which makes it very hard to understand what they're signing. Third, there's a barrier between you and the speaker; it's very hard to participate in discussions" (Mandula, 1987, p. 7). At one scientific meeting Dillehay attended, the lecturer did not want the interpreter there and asked the interpreter to leave. After Dillehay persisted, the scientist grudgingly let the interpreter stay. During a conference held by a very prominent association which I attended with several other deaf scientists in 1988, one presenter was speaking at such a rapid rate that the translation to American Sign Language (ASL) by the most expert of professional interpreters was made very difficult. After being asked twice very politely by the interpreter to slow down the pace of the presentation a little, something the hearing scientists would likely have also appreciated, the presenter curtly suggested that perhaps the deaf scientists should just sit and read the synopsis he had handed out. Schein (1975) also described how seldom at professional meetings a hearing colleague will show a deaf scientist the courtesy of inquiring if he or she wishes to contribute to the discussion.

The importance of scientific societies taking increasing responsibility to provide access to scientists with disabilities cannot be overemphasized. I have shown how scientific societies and academies played significant roles during the Enlightenment in bringing recognition to the need for improved educational opportunities for deaf children. In Europe, interest in deafness and its relationship to the study of language (both signed and written/spoken language) stimulated early thinkers in the societies and frequently led them into associations with instructors and school programs for deaf children. For those few deaf scientists that there were, these societies also provided opportunities for communicating, primarily through writing, their progress in scientific investigations.

Also, as previously discussed, the "deaf experience" is also interwoven throughout the histories of national and local scientific societies in the nineteenth and early twentieth centuries in America. In the National Academy of Sciences, the profoundly deaf paleobotanist Leo Lesquereux was the first elected member, and the physicist Frederick Barnard and astronomer Robert Aitken were active members. Gideon E. Moore worked with the NAS to study the production of sugar from sorghum after the Civil War. Barnard was also the president of the American Association for the Advancement of Science, and in the history of the AAAS can be found the experiences of deaf people such as the botanist Thomas Meehan who lectured on his work with Darwin, and the physician Robert Farquharson's presentation on the Mound Builders of Iowa. Farquharson helped to establish the Iowa Academy of Sciences. Fielding Bradford Meek, like his contemporary Lesquereux, was honored with membership in many societies, yet, primarily because of deafness, avoided their meetings. Through correspondence with Joseph Henry, James H. Logan used the Smithsonian Institution as his primary source for information about specimens and in evaluating his ideas which led to a patented improvement for a microscope, and Gerald M. McCarthy donated to the Smithsonian Institution many of his plant specimens obtained during his botanical travels. Logan helped to found the Iron City Microscopical Society in Pennsylvania. Over the past fifty years, and particularly in the last two decades, participation of scientists with disabilties in scientific societies has grown. The question thus remains: How much more would these societies be enriched if they were made accessible to the deaf scientist?

This issue of access is of paramount importance and it behooves us to address it as a priority in light of the predicted shortage of scientists and engineers in the United States. While the twentieth century has seen great advances in technology, education, and federal legislation, every indication is that underrepresentation of people with disabilities in science remains a serious problem despite worthwhile attempts by the National Science Foundation, U.S. Department of Education, and other agencies to address the issues. Of paramount importance is continued expanded federal support, particularly in enhancing precollege and undergraduate educational opportunities.

The Need for Continued Activism

In an article titled "Disabled Women: At the Bottom of the Work Heap," Susan June McLain and Carol O. Perkins (1990) described how women with disabilities must fight a battle within themselves to gain a feeling of competence. As children, they were not encouraged to develop the traits needed for success in the workforce. Fear of success, a lack of role models, and inadequate career guidance are also factors. The authors quote Charlotte Toews in

Resources for Feminist Research: "As women have fought patriarchy, the disabled have fought paternalism—regardless of its benevolence. As women have marched to take back the night, the disabled have demonstrated to gain the right to public access" (McLain & Perkins, 1990, p. 54). To change the status quo, McLain and Perkins encourage women with disabilities to become involved in both the women's movement and the movement for persons with disabilities, and seek out support groups to help them further their career.

This is also very good advice for the deaf scientist, male or female. Becoming involved in the Deaf community and seeking out the support of others who share similar experiences has long been a successful strategy. The NAD, in particular, was the first organization of people with a disability in America established to provide such support. Members of NAD have been effective leaders in the effort to improve educational and occupational conditions for deaf people and the NAD has provided a channel for deaf scientists and engineers to take important roles as community leaders. As previously stated (Chapter 2, "Gallaudet College's Impact on Science as a Profession for Deaf People"), the chemist George T. Dougherty was a founder and first secretary of NAD. Frederick C. Schreiber, whose training at Gallaudet College was in chemistry, became the NAD's first executive secretary in 1966, an office in which he served until his death in 1979. Another chemist, Robert O. Lankenau, was elected president for 1968–1972, and T. Alan Hurwitz, a former associate electronics engineer, was NAD president from 1982 to 1984. Through the years, deaf scientists and engineers collaborated with other deaf leaders in effecting positive change in employment opportunities for deaf people, including Benjamin M. Schowe, Sr., a labor economist specialist who helped to develop numerous informational publications on deaf employees; Boyce R. Williams, who, in his thirty-eight years in the Office of Vocational Rehabilitation, promoted cooperative relationships among state and federal rehabilitation programs and organizations serving deaf adults; and Robert R. Davila, who in 1989, became assistant secretary in the U. S. Department of Education's Office of Special Education and Rehabilitative Services, the highest-ranking deaf person in the history of the federal government.

The active role NAD members have taken in implementing state and federal statutes include efforts which have led to amendments to many legislative acts in vocational rehabilitation, social security, and education, all beneficial in opening up new avenues for deaf people.

The positive and fruitful experiences of deaf scientists and their associates point to the need for continued self-advocacy and activism. While the nature of scientific study and deafness has led many to a more contemplative way of life, human progress is not made in the laboratory alone. Deaf scientists who have been active in pursuit of their own and others' rights have contributed to a legacy in which opportunities in science may become the rule, rather than the notable exception.

Epilogue

One day in 1988 I took a train from Leeds in northern England to visit the beautiful town of York, where, two hundred years earlier, young John Goodricke had peered up at the sky through the windows of the Treasurer's House. Walking from the train station in York along the ancient Roman walls, I paused to reflect on what the town must have looked like in Goodricke's time. The Treasurer's House still stood by the York Minster, and, after musing for a while about Goodricke's family, I entered the splendid cathedral which had been built in the twelfth century. Few astronomers in history, I imagined, have had a more beautiful "fixing point" for their stellar measurements.

I lived this book as I wrote it. In the towns and libraries I visited, in the letters I penned and received, and in the many volumes I read, history became a part of my own life. In this sense I hope that through these efforts of mine history will also become more meaningful to many deaf students. I remember early one morning in December 1987 when I crossed the Neckar River and hiked to the top of Heiligenberg ("Holy Mountain"), past an ancient monastery, and sat alone on the stone steps of a Roman amphitheater overlooking Heidelberg, Germany, pondering whether Gideon E. Moore had ever done the same while he had lived there. Moore, too, would have spent much time looking across the Neckar at the wonderful medieval castle built upon the hill. A little over a century earlier, he had studied under the great chemist Robert Wilhelm Bunsen. I daydreamed about sitting in Bunsen's class as I sat in the early morning mist in this place of solitude and, frustrated with my own efforts to learn the German language, I wondered mostly about how the deaf American chemist had managed to communicate well enough to earn his Ph.D. *summa cum laude*.

The following summer, I traveled along the Missouri River on my way to the Badlands. This time, I dreamed of being on the steamboat with the geologist Fielding Bradford Meek 133 years earlier, observing the Native Americans along the shore. I pictured myself floating past the spreading elms and stopping for the night to gather firewood, and specimens of fossils and insects, while slowly making my way toward Fort Pierre, and then onto Mauvaise Terres.

I remember that special day in England when I had gained permission to browse through the rare book collection in the archives of the University of Leeds. I stood among the large, old volumes with their titles in Latin, Greek, German, and other languages. There were early editions of scientific classics, some faded and tattered, others still reflecting bright colors which, like their contents, would live on in the minds and hearts of humankind for years to come. They all waited to be read, and I did not know where to start. Hours later I came upon one of my most exciting findings, the "Éloge de M. Amontons" ("Obituary for Monsieur Amontons") published in 1705 in the *Histoire de L'Academie Royale*. The reports of a profoundly deaf physicist demonstrating the results of his experimental research on heat and temperature in the chambers of the French Academy of Sciences in the late seventeenth century left me completely satisfied with this one discovery, all I had to account for after examining several hundred manuscripts during that two-day experience with antiquity.

Searching out the lives and work of deaf scientists also proved rewarding in another way. I found myself visiting classes in schools for deaf students, telling the story about the deaf chemist Ekeberg, who had discovered one of the elements. Granted, I was very much a novice storyteller, but it was thrilling to see the change in the way the deaf students began to look at the periodic chart, now that they had a personal way to identify with it. This was, after all, the *raison d'être* for my work and I was finally beginning to see the fruits of my labor in the classroom.

Some fascinating stories can be told to deaf students to stimulate interest in mathematics and science, although they do not always have immediate practical value. One such story is the humorous and strange coincidence involving the writer Jonathan Swift. Swift, whose deafness began when he was about twenty years old, wrote *Gulliver's Travels*, a tale of the adventures of Captain Lemuel Gulliver in which astronomers of the Isle of Laputa "discovered two lesser Stars or satellites, which revolve about Mars" (Swift, 1959, pp. 160–161). In his book *The Star Lovers*, Richardson (1967) compared the Laputan astronomers' data for the orbits of the two satellites with those actually observed for Phobos and Deimos 150 years later by the astronomer Asaph Hall. "Although there is considerable gap between fact and fiction," Richardson wrote, "the agreement is nevertheless remarkably close . . . so close, in fact, that some are convinced that Swift had access to a secret source of information not available to the rest of us" (p. 154). Of special interest to deaf people who have followed Gallaudet University's history like a family is the fact that Asaph Hall was a hearing man whose son, Percival, later became its second president after Edward Miner Gallaudet retired. Asaph Hall reached eminence at the U.S. Naval Observatory for his discovery of the two potato-shaped moons of Mars in August of 1877. Unhappy that his observations of Mars through the great equatorial telescope in Washington, D.C., had yielded nothing, Hall had given up and gone home on that fateful night. His wife then encouraged him to try one more time and Hall returned to his work. He spotted one moon at 2:30 A.M. on August 11, the other six days later.

Mars is not an easy planet for astronomers to study. The diffuse light surrounding it makes it only dimly visible. In personal correspondence, Asaph

Hall's grandson, Jonathan Hall, described to me how this problem was circumvented. Asaph Hall blocked out much of the light by keeping Mars outside his field of view, just as one would reduce the glare from the sun by shading the eyes with a hand. Astronomers have since failed many times to find the moons, even with much more powerful telescopes. The fact that no astronomer had ever seen one of these moons of Mars until Hall's discovery adds to the mystery of Jonathan Swift's writing. "The innermost [moon]," wrote Swift, "is distant from the centre of the primary planet exactly three of his diameters, and the outermost five; the former revolves in the space of ten hours, and the latter in twenty-one and a half; so that the squares of their periodical times are very near in the same proportion with the cubes of their distance from the centre of Mars: which evidently shews them to be governed by the same law of gravitation that influences the other heavenly bodies" (p. 161).

Herein mathematics enters the story. Some questioned Swift's capability for any kind of mathematical rigor, leading some to believe he had assistance. Yet, Swift's dislike for mathematics did not reflect his ability in that subject. One day, Swift's friend Dr. Sheridan, in settling a dispute with him that the science was not difficult to master, gave him a challenging problem to solve. Swift was left alone for half an hour and, at the end of that time, the cry of Archimedes could be heard when he had solved it accurately.

My point in telling this story at the end of my own account of literal and metaphoric travel is this: While Swift's deafness and the fact that Hall's son became the president of Gallaudet College had no bearing on the strange coincidence of Gulliver's moons of Mars, they are illustrative of how facts and anecdotes may be transformed without fabrication in a manner that may captivate the interest of both deaf and hearing children. Such stories, and the actual life experiences of deaf men and women in science, add a missing dimension to Deaf Studies and provide a unique perspective on history.

Afterword

SILENCE OF THE SPHERES

I sit once again on my favorite knoll and watch the stars glitter in the tranquil night sky. Far beyond our solar system, partly "deaf" with antenna reception problems, Voyager carries a gold-plated copper disk tucked into a silvery aluminum cover with recorded greetings in nearly sixty languages. The phonograph is, of course, the invention of Thomas Alva Edison, and 1977, the year of the Voyager launch, was the hundredth anniversary of its development. These were appropriate tributes to the deaf inventor. The final musical piece in Voyager's collection is the first two bars of Beethoven's Cavatina from the String Quartet No. 13 in B-flat, Opus 130, composed two years before his death, and at a time when the composer was completely deaf. Beethoven's music played on Edison's phonograph may some day bridge the spheres of silence and sound.

In 1914, Rena Albertyn Smith penned an essay entitled "The Veil of Silence," in which she discussed the spirit of music found in the vagrant wind "whispering its fancy"; in dancing, in the hushed waters where changing colors play, in drama and poetry, and in the emotions of facial and bodily expression. "We may not hear poetry spoken by the musical voices of men," she wrote, "but our enjoyment is nevertheless exquisite, for our speakers are the voices of visions and our ears are spirit ears." Smith captured the essence of her writing in her concluding paragraph:

> In silent surroundings the thought may arise that deafness is a dreary stillness, and that the deaf find terror in the silence, whereas it is as probable that the loneliness which the man of hearing experiences when he is where sound sleeps is caused by the lack of human fellowship, while contact with the silent immensity of the infinite produces a wonder akin to awe. And awe, indeed, there is, for the voice of the soul speaks when sound is absent, directing the light of man's thoughts within. As the soul whispers in still moments, so does it in that longer silence which is

> deafness. But the deaf person, grown familiar with silence, has recovered from the disquiet which this enforced communion arouses; and within his own soul he finds the tie binding him to the souls of others—a tie that is often twisted and torn, but never quite severable between men, though men themselves may turn into dust. And the extent of this spiritual kinship realized, there is an hour of renascence in which he recognizes that in the sphere of silence he may not have lost part of his birthright, but come into a fuller heritage. (p. 205)

It is about this "fuller heritage" that I have written in this book. And, as some deaf people may seek the "sphere of sound," the world of the hearing dominated by music and spoken language, the "sphere of silence" is sometimes desired by hearing men and women as well. It is the stillness of the countryside, the quiet of the monastery, and the solitude that poets and writers and champions of science have long sought for undivided thought.

Pythagoras's conception of the "music of the spheres" eventually became abstract, and evolved to a more scientific representation of celestial motion which provided the foundation for astronomical thinking until Kepler's revolutionary work. We frequently find references to this notion of "music of the spheres" in the literature of medieval and modern writers. Wilczek and Devine (1987) asked the question whether there truly is a "music of the spheres." They argued that this "marvelous dream" of Pythagoras is in fact closely realized in the physical world: "The spheres, however, are not planets but electrons and atomic nuclei, and the music they emit is not in sound but in light" (p. 14). They described how our sensory systems process light and sound. We cannot detect pure colors in light as the ear can recognize individual tones in a musical chord, but light can be separated into its component colors. "It's a truly wonderful thing," they explained, "that it's possible to recover the lost information. It is as if we were all tone-deaf and had suddenly discovered a way to appreciate music" (p. 14). We experience this "music" through our eyes as we study the flame spectra of chemicals, each element producing its own chord. "If our eyes were more perfect," they wrote, "we would see the atoms sing" (p. 15).

The astronomer Zdenek Kopal (1986) also once described a visual metaphor of the "music of the spheres" which, he explained, has been played for us by the eclipsing binary star systems in the course of their evolution. As one star eclipses the other, changes in light can be observed from Earth, and these changes carry much information (p. 426).

Binary stars. Companion stars—silent sensations which follow the brightening and darkening of two independent, interdependent worlds. It is an ethereal music that I enjoy as I sit here tonight, a music that is, as the great astronomer Johannes Kepler described in 1619, "perceived by the intellect, not by the ear."

Hearing people *were* the original composers of deaf people's lives. But, very gradually, deaf men and women, once relegated to playing a minor accompaniment in the history of science, emerged from an eclipse of ignorance and prejudice and entered the light of reason. Sharing energy, and often now directing it, they work with their hearing companions to sharpen all of our sensations, and to further human progress.

Notes

Preface

1. Craters of the moon named in honor of deaf men and women mentioned in this book include, on the near side, Crater Cannon, which is named for Annie Jump Cannon. On the far side, there are craters named for Konstantin Eduardovich Tsiolkovsky, Henrietta Swan Leavitt, Oliver Heaviside, Charles de la Condamine, and Robert Grant Aitken.

Introduction

1. The Little Paper Family began in 1849 with the publication of a periodical, *The Deaf Mute*, at the North Carolina School for the Deaf and Blind. Other schools for deaf children across the country soon followed with their own newspapers or journals. The periodicals were used to teach printing, a trade commonly pursued by deaf persons in the nineteenth century. The papers also served well in sharing information about the schools and in presenting issues important to deaf people. As John V. Van Cleve (1987) explained, the "Little Paper Family . . . is more than an artifact of the deaf community. It was an institution that helped deaf Americans forge and maintain their identity; it provided a training ground for leaders; and it led to employment for deaf people" ("The Little Paper Family," *Gallaudet encyclopedia of deaf people and deafness*, Volume 2, pp. 193–195). Many Little Papers are available at the library at Gallaudet University.

2. Milton E. Larson presents a brief summary in the *Phi Delta Kappan* (LIV, 6, 1973, p. 374) of how many famous people have faced such attitude barriers early in their lives.

3. Undated newspaper clipping. Gallaudet University Archives Biographical File "George K. Andree."

4. H. J. Conn, "A religious scientist at the turn of the century: Herbert William Conn of Wesleyan University." Manuscript at American Society for Microbiology Archives, pp. XI–8–XI–9.

5. The isolating conditions of deafness are revealed in many forms through history. One report which appeared in the September 17, 1919, issue of *The Westminster Evening Institute*, for example, described a club formed for deafened ex-

servicemen in England after World War I "to promote the social welfare of a very considerable body of deserving men, when they will have an opportunity of indulging in sports, games, and other recreations according to their individual tastes without the embarrassments which they would be inclined to feel when mixing with others who enjoy their faculties unimpaired" (p. 625).

Chapter 1

1. The ancient Greeks viewed deafness as a disgrace. The Hebrews placed restrictions on deaf people in regard to legal transactions, ownership of property, and marriage, and the Assyrians thought diseases of the ear were caused by the hand of a ghost seizing a person. Even as late as the eighteenth century barbarous treatment of deaf people continued in some countries. In primitive tribes, the ill and "handicapped" frequently committed suicide. Depending on the nature of the "handicap," attitudes within a single tribe varied. The Bayaka in the Congo, for example, respected the blind but derided deaf people (H. E. Sigerist [1955], *A history of medicine*, New York: Oxford University Press, p. 157).

To illustrate the general attitude about instructing deaf people that was prevalent in ancient times, most historians cite the works of the philosophers Aristotle and Lucretius. Fallacious beliefs about deafness and learning persisted for centuries and explain the appalling obscurity into which deaf people were thrown for fifteen hundred years following the time of Christ. Aristotle's writings covered a wide range of subjects, including astronomy, psychology, politics, ethics, metaphysics, biology, and art. It was in the realm of biology, his chief interest, that Aristotle wrote on deafness. Translations and mistranslations of Aristotle's writing in *History of Animals* led to a view of deaf persons as "senseless" and "incapable of reason" (Bender, 1981, p. 21). In his poetry, the Roman philosopher Titus Lucretius Carus expressed an obviously oppressive view on the education of deaf people: "To instruct the deaf no art could ever reach, No care improve them and no wisdom teach." *De Rerum Natura* (*On the Nature of Things*) was not a scientific work, but it exercised great influence on scientists of the early modern period. It had much less effect on scholars of the Middle Ages. For a while it appears to have been lost. But when a copy was later found, Gutenberg's printing press helped it to quickly fall into the hands of modern scholars and, with Lucretius' philosophical account of the universe, the mistaken belief that deaf persons could not be taught was continued.

2. Little is known about John the Deaf. Andrew Fleming West in his book *Alcuin and the Rise of the Christian Schools* (1892) wrote that "John the Deaf . . . instructed Roscellinus of Chartres." He was also called John the Physician, of Chartres. F. DeLand (1924) wrote that there is "such a scarcity of information on the deaf scholar and physician that the task of tracing him appears hopeless." Yet DeLand's assumption that the tenth-century physician was deaf is difficult to accept. As I will show in this book, even late-deafened physicians or those with mild hearing losses have been discouraged from continuing their medical practice, and most contemporary deaf physicians will attest to the stigma associated with deafness and the prejudice they have experienced in medical school and in their careers.

Hoping that a half century of scholarly work by experts in medieval literature may have uncovered new information since the time of DeLand's publication, I embarked on my own investigation of John the Deaf. With the assistance of Alan Lupack at the University of Rochester's Robins Medieval Library, I found that John the Deaf is recognized by scholars as the founder of nominalism, the branch of philosophy dealing with the naming of universals. But this, too, was clouded in obscurity and fragmented historical records. A second "John the Deaf" was then found. Also known as John of Paris, John Quidort, Jean Le Sourd, and Johannes de Soardis, he was born around 1255 in Paris and died on September 22, 1306, in Bordeaux, Gascony. He was a Dominican monk known for his work in philosophy, theology, and political theory. Although it was unusual that he had received his master's degree when he was over fifty years old, especially in light of what we know of university practice during this period in history, scholars who studied his life have not conjectured that a hearing loss may have delayed his education.

3. Rudolph Agricola (1443–1485), born in Groningen, the Netherlands, was a talented humanist, philosopher, theologian, musician, and painter, and was fluent in Greek, Hebrew, and Latin. In *De Inventione Dialectica*, published posthumously in 1528, Agricola mentioned the deaf man with the ability to write. In *Paralipomenon*, Jerome Cardan (1501–1576) described the potential of deaf persons to "hear" by reading and to "speak" by writing. In addition to Cardan's son's being deaf in one ear, Cardan himself was a stutterer, and this may have led him to take a special interest in the subject of speech and hearing. Cardan was also a friend of Leonardo da Vinci and was knowledgeable about his work.

4. Wright (1969) presents a prose translation of the epitaph written for the deaf painter Juan Navarette ("El Mudo"), the Spanish Titian, by the prolific dramaticist Lope de Vega: "Heaven denied me speech, that by my understanding I might give greater feeling to the things which I painted; and such great life did I give them with my skilful pencil, that as I could not speak I made them speak for me" (p. 141). Not far from Navarette lived the "Prince of French Poets," Pierre de Ronsard (1524–1585), who spoke to the world through his pen. Stimulated by the Italian Renaissance, he later rose above the arid imitation of classical work of his contemporary poet-scholars to produce lyrics that greatly influenced poetry in France. A third deaf man who reached eminence, Joachim Du Bellay (1512–1560), a fellow Frenchman, used the medium of poetry to express his emotions about deafness. His "Hymn to Deafness" was a tribute to Ronsard.

5. Despite Beethoven's progressive deafness, he continued to triumph with his musical masterpieces. Around 1814, an inventor provided him with an ear trumpet, but within three years he was almost completely deaf. His correspondence reveals the torment caused by his acquired hearing loss, and alludes particularly to the sense of social isolation.

6. Socrates to Hermogene: "But answer me this question: If we had neither voice nor tongue, and yet wished to manifest things to one another, should we not, like those which are at present mute, endeavor to signify our meaning by the hands, head and other parts of the body?" Plato, *The Cratylus, Phaedo, Parmenides, and Timaeus of Plato*. (Translated from the Greek by Thomas Taylor. London: Printed for Benjamin and John White, M.DCC.XCIII [1793].)

7. By the time his *Essay Concerning Human Understanding* had gone through its fourth edition in 1704, John Locke also had experienced deafness. Dewhurst's (1963) description of Locke's physicians' attempts to cure him over a period of three years provides a glimpse into the state of medical science in the early eighteenth century: "It began with a chill which brought on a severe bout of asthma and bronchitis. . . . a few weeks later, when he suffered a violent earache, Dr. Alexander Geikie treated him with an application of a large roast onion wrapped in Colewart leaf, and made into a poultice with the addition of herbs. James Tyrrell also advised an onion poultice, and the application of 'woman's milk warmed' with juice of rue, whilst Dr. Guide suggested bread hot from the oven soaked in Eau de Vie, and should this fail to cause suppuration, then 'oil of worms' in which you have boiled snails and woodlice distilled and then dropped into the ear with a slice of onion or garlic" (pp. 287–288). It is not known if any of these remedies relieved Locke's pain, but at least one of them made the philosopher even more deaf. Locke carried with him a silver ear trumpet. Other

solutions were attempted through time until, one day, an internal abscess burst spontaneously.

8. While John Conrad Amman's well-known *A Dissertation on Speech* was still in press in 1700, he received a letter from the celebrated John Wallis, and through their subsequent exchange Amman and Wallis shared views on their work with deaf children. In January, 1700, for example, Amman wrote to Wallis that "if I were addressing a man less courteous and learned than yourself I should have to make many apologies for the liberties I have taken. . . . I earnestly entreat you if you find any thing calling for your animadversion, candidly to tell me of it, and wherever I seem to you to be in error to correct me as a friend" (Amman, 1873, p. xxviii). Wallis's correspondence with Amman also led to Amman's book being translated into English by the physician Daniel Foot. Pioneers in the education of deaf children in eighteenth-century France, including Jacobo Pereire, Azy D'Etavigny, and the Abbé de l'Epée, read Amman's work.

9. Special Collection 371 92092, f.B4, as described in R. V. Jones and W. D. M. Paton (Eds.), *Notes and records of the Royal Society of London* (1974–75), p. 58.

10. Taken from the most recent edition (Johnson, 1990, p. 126).

11. *Miscellaneous papers respecting deaf-mutes in France, Germany, Italy, Austria-Hungary, Belgium, Switzerland, Holland, and the United States.* Presented to the House of Lords by the Command of Her Majesty in pursuance of their address. August 13, 1885.

12. Alexander Small, Benjamin Franklin's acquaintance, published various reports in the *Philosophical Transactions* of the Royal Society, including one on ventilation. He wrote to Franklin in 1777, requesting that he put the finishing touches on his report. In subsequent years, Small corresponded with Franklin about a serious case of gout which Franklin had tried to cure himself. In one letter to Small, dated February 17, 1789, Franklin wrote that "the deafness you complain of gives me concern, as if great it must diminish considerably your pleasure in conversation."

Lawrence's deafness began in middle age. He became a fellow of the Royal College of Physicians of London in 1744 and was elected president in 1767. He shared experiences with deafness with his close friend, the writer Samuel Johnson.

Aloys Weissenbach (1766–1821) completed his studies in 1788 at the Josephs Academy in Vienna. After years as a surgeon in the field army, where he advanced to senior field surgeon, he held a teaching position at a university in Salzburg. Weissenbach was also a talented poet and writer. His tragedy *Der Brautkranz* in iambics, five acts, was produced January 14, 1809. Beethoven decided to commemorate the gathering of royal guests in Vienna for the Congress by presenting the deaf physician Weissenbach's poem "The Glorious Moment" in a musical setting. The cantata was a difficult challenge to set to music, and the two of them labored to revise and polish it. With the assistance of Karl Bernard, it was completed in time and was presented with several other pieces at an evening concert. It is interesting to imagine what thoughts must have drifted through the minds of these deafened gentlemen as they discussed the cantata under such conditions.

13. Samuel Latham Mitchill, founder and editor of the *New York Medical Repository*, published in early volumes of this journal numerous scholarly writings on geological matters. He had also published the writing of Francis Green on the early education of deaf pupils. In 1818, Mitchill authored *A Discourse Pronounced by Request of the Society for Instructing the Deaf and Dumb at the City Hall in the City of New York.* He was a vice president of the Society and in this work he described the efforts of Thomas Hopkins Gallaudet and Laurent Clerc in educating deaf students, as well as those of François Gard in France, with whom he had been corresponding.

14. Thomas Hopkins Gallaudet and Laurent Clerc also met with Native Americans of different tribes shortly after the American Asylum was established in

Hartford, Connecticut. Like Akerly and Long, they found some similarities in the signs used by the Native Americans and American Sign Language, with its predominantly French origin.

Chapter 2

I would like to thank Phyllis Harper-Bardach and Pamela Culver for sharing information about R. J. Farquharson for this chapter.

1. Bunsen must have been no stranger to deafness. In addition to the American chemist Gideon E. Moore, Bunsen's former student Baron Carl Auer von Welsbach (1858–1929) also experienced deafness. In a biography in a German chemical journal, J. D'Ans (1931) explained that Auer's progressive hearing loss was possibly initially due to artillery fire. D'Ans wrote, "Auer did not belong to those of a social disposition. As he became increasingly hard of hearing, he was driven even more into solitude" (p. 61). His dog, "Buzi," was with him day and night. Always observant, he would jump on the deaf chemist to alert him of a person or animal approaching from behind, and, in the laboratory, would not allow anyone but Welsbach to touch even a piece of paper. Among Welsbach's inventions was the gas mantle, which competed successfully with the incandescent bulb for some time. The mantle, patented in 1885, was developed when he discovered that various compounds associated with the rare metals of the thorium group become luminous when held over a flame. It consisted of threads of cotton impregnated with solutions of the nitrates of thorium and cerium. The oxides of these two metals formed by igniting the solutions emitted luminous radiation when heated by a gas flame. When Welsbach visited his "old master" Bunsen in Heidelberg and demonstrated the mantle, Bunsen "shook his head and said that it seemed impossible to him that a coherent mass of oxide could be obtained in that manner" (von Welsbach, 1902, p. 254). Welsbach's gas lighting held its own against electricity until World War I. Today, the most popular modern use is for portable lamps.

2. The predominant areas of specialization of these doctoral degrees are in the medical sciences (77), chemistry (25), biological sciences (14), mathematics/computer science (11), engineering (9), and physics/astronomy (7). The general distribution of the larger sample, which includes master's degrees and bachelor's degrees, is as follows: (a) for deaf men, the most common field was chemistry (137), followed by mathematics/computer science (97), engineering (76), medical sciences (68), biology (67), and physics/astronomy (30), and (b) for deaf women, the most common area of concentration was biology (37), followed by medical sciences (28), mathematics/computer science (25), and chemistry (14).

3. Gallaudet University Archives Biographical File "Ide L. Kinney."

4. The seventeenth-century physicist Guillaume Amontons might be credited as having been the first deaf person to make a meaningful contribution to geology. In one of the earliest discussions of earthquakes, Benjamin Franklin wrote in 1788 to the Abbé Giraud-Soulavie, suggesting that the earth may not be solid to the core. In this letter, Franklin mentioned the work of Amontons, who "calculated, that its density [increased] as it approached the center" (Merrill, 1924, p. 13). Franklin, however, was likely not aware of Amontons' deafness. Amontons' seventeenth-century contributions to this field of study are rarely mentioned. Soulavie, credited with having "planted the seeds of stratigraphic geology" in his 1779 paper, was in turn eclipsed by such successors as Cuvier, Brongniart, and Desmarest.

5. The *Journal of Botany* (*55*, 1917, p. 145) reported that the British paleobotanist Clement Reid (1853–1916) was deafened by scarlet fever in childhood, although he apparently regained most of his hearing after being deaf "for some

years." Reid worked with Alfred Nathorst in excursions in Yorkshire. He published many reports on glacial deposits, fossil flora of submerged forests, and the natural history of isolated ponds. Reid discovered hundreds of species of flora, and his investigations of the complicated Pliocene and Pleistocene deposits led to a special interest in the discovery and identification of the seeds of plants. His mother was a niece of the noted electrical scientist Michael Faraday, and the "influence of the great scientific spirit of Faraday permeated the whole surrounding of his childhood, and gave encouragement to the natural bent of his mind" (p. 145). Thus, Reid was instilled with his mother's love of nature, and this helped, drawing him to a career in which he won many honors, including membership in England's Royal Society.

6. The land for the Columbia Institution was provided by Amos Kendall, a journalist, businessman, and politician. Kendall's editorial work had earned him the nickname "the Robespierre of America." He was postmaster general under President Andrew Jackson and the most influential member of his "kitchen cabinet." Early in his life he had an inclination toward science and invented a pump. He became rich as the business manager for Morse. Incidentally, after Morse's first wife's death, he fell in love with and married a deaf woman, Sarah Griswold. In *The American Leonardo: A Life of Samuel F. B. Morse* (1969), Mabee wrote: "Whether because of her poverty or her difficulty of hearing and speech, she quickly became his attentive helper without presuming, so far as abundant correspondence indicates, to be his legal or financial adviser. She showed the dependence he had expected and desired" (p. 306). Dependent or not, one of the first joys Sarah experienced in her marriage to Morse was a trip to Frankfort, Kentucky soon after the wedding to attend one of the many patent lawsuits that plagued the life of her husband. Samuel and Sarah Morse failed by one year to celebrate their silver wedding anniversary.

7. President Abraham Lincoln signed the charter authorizing the Columbia Institution to grant college degrees on April 8, 1864. But the Civil War was raging and the president's signing of the charter for the first college for deaf students was little cause for excitement in the Deaf community. All over the nation teachers and administrators had left their positions in schools for deaf pupils to take up arms. During the conflict, the Columbia Institution buildings housed soldiers from Rhode Island and Pennsylvania. Its physician, Alexander Y. P. Garnett, resigned to take up the position of surgeon general of the Confederate Army and personal physician to Jefferson Davis.

8. Greely himself published on the subject of sign language in the *American Annals of the Deaf*, and he wrote an article about the college for deaf students in *Review of Reviews*. Henry had several decades earlier discovered the principle of electromagnetic induction, which eventually led to the development of the telegraph. This he did independently of Michael Faraday at the Royal Institution in London and the German scientist H.F.E. Lenz working in Russia; Henry's onerous teaching and administrative duties at the Albany Academy, however, delayed his publication of the discoveries until 1830.

9. Henry's personal library included copies of Charles Edward Orpen's *Anecdotes and Annals of the Deaf and Dumb* (1836) and the *Fifteenth Annual Report of the Directors of the New York Institution for the Instruction of the Deaf and Dumb* (1834). He and other scientists were very interested in Laura Bridgman, the deaf-blind girl tutored by Samuel Gridley Howe. *On the Vocal Sounds of Laura Bridgman* (1850) was published by the Smithsonian Institution.

10. Personal Communication, Anne Enegelhart, March 3, 1988.

11. Personal Communication, Barbara L. Welther, November 14, 1989.

12. Personal Communication, Mary B. Bowling, July 29, 1987. Appreciation is also due Reese V. Jenkins, director and editor, Thomas A. Edison Papers, Rutgers, The State University of New Jersey. Personal communication, June 17, 1987.

13. *Our Deaf and Dumb*, *2*, June 1895.

14. *The Western Pennsylvanian*, *5*, 2, 1896, p. 4.

15. Deaf inventors had already demonstrated their talents before the Civil War. Among them was Robert William Thomson, who invented a rubber tire for the bicycle in 1845. In 1855 two deaf Frenchman, M. Maloisel and M. Richardin, were honored with distinguished prizes at the Paris industrial exhibition. Maloisel, a former pupil of the Paris institution, won his prize for a machine that executed sculpture. Richardin, from the institution at Nancy, invented a polishing machine for daguerreotype plates (*AAD*, *9*[3], 1857, pp. 174–168).

Chapter 3

1. Over the years an astonishing variety of "cures" for deafness has been reported in the literature, ranging from ground worms boiled in goose grease to the "night chant" of the Navaho Indians, a nine-day affair in which different chants were offered each day for the purpose of curing a deafened member of the tribe. Some "cures" were offered through the work of scientific societies, as in the following report found in *The Lancet* (June 10, 1848): "Dr. Baudelocque has presented to the French Academy of Sciences, a boy seven years of age, born deaf and dumb [sic], whom he intends to treat for this defect. He requests that the Academy will keep an eye upon the child for one year, considering that the patient being examined from time to time during this twelve months, will conduce to the settlement of the question of the curability of this affection" (p. 652). The fate of the deaf boy is not known, but twelve years later a warning appeared in the same medical journal that the administration of chloroform as a cure for deaf soldiers and a young deaf boy was "very dangerous." Baudeloque is reported to have presented the subjects of his investigations to the Academy in Paris. The dangers were, of course, not limited to deaf persons. In 1851, Joseph Toynbee, an assistant curator to Professor Owen at the Royal College of Surgeons, believing that his own tinnitus (head noises) might be reduced, subjected himself to chloroform inhalation and was found dead in his consulting room with his notes and equipment by his side.

Later, the era of flight ushered in predictably human, predictably mad "cures" as well. Following reports by several individuals who had claimed nose dives had cured their deafness, a number of deaf people and their pilots were killed in attempts to regain their hearing.

In *The Silent Worker* (*34*[2], 1921, p. 60) can be found an advertisement encouraging deaf people to send fifteen cents to the Propaganda Department of the *Journal of American Medical Association* in Chicago for *Deafness Cures*, a pamphlet of forty-four pages exposing fake instruments and quack medicines for curing deafness. The editor of *The Silent Worker* commended the AMA for its "fearless fight against quackery."

2. Disenchanted with the deprivation of basic rights and privileges routinely accorded hearing persons of the time, one deaf individual in the nineteenth century, James J. Flournoy, issued a circular to the deaf people in the United States and Europe, calling for the establishment of a commonwealth for deaf people. Flournoy pleaded for a state populated only by deaf persons to be founded with its own congressional representative. Through the remainder of the century, publications of the Deaf community continued to report interest in establishing "colonies" of deaf people, as in the case of about one hundred deaf people in a "resort club" in Kansas in 1893, and as evidenced by a banking firm in northern Wisconsin which offered "very advantageous" terms for selling 15,000 acres to "worthy deaf persons" for the purpose of

developing a colony. The Deaf community also reported on the high concentration of deaf people at Martha's Vineyard as a "Deaf-Mute Colony."

3. See R. Winefield (1987), *Never the Twain Shall Meet: The Communications Debate* (Washington, DC: Gallaudet University Press) for a more thorough analysis of the historical controversy over sign language and "oral" communication.

4. Robert Grant Aitken's "Record of Family Traits" (Eugenics Record Office, Cold Spring Harbor, Long Island, New York), dated February 13, 1925. National Academy of Sciences Archives.

5. R. Pearl to E. B. Wilson, March 9, 1923; E. B. Wilson to R. Pearl, March 9, 1923; R. Pearl to E. B. Wilson, March 12, 1923. Washington, DC: National Academy of Sciences Archives.

Chapter 4

1. A copy of this letter may be found in the Gallaudet University Archives Biographical File "Gerald M. McCarthy." Compare this lack of deaf teachers at Gallaudet during the early years with the fact that, by 1873, 73.1 percent of 1553 teachers in institutions for women were female (Rossiter, 1982, p. 9).

2. Such a low number precludes any effective labor market analysis of deaf people in science. In addition to the deaf chemists mentioned in this chapter, others working at the end of World War I included Carl Bohner (metallurgist for the Pennsylvania Railroad), Charles L. Clark (chemist for the Delaware and Hudson Railroad), Archibald Wright and Lawrence E. Johnson (both chemists at the National Health and Welfare Department in Canada), Ralph Decker (chemist at Arkansas Zinc Company), Wallace K. Gibson (chemist at Trinity Portland Cement Company in Dallas), Gerald Ferguson (chemist at the Bureau of Engraving and Printing in Washington, D.C.), Wallace D. Edington (chemist at the Department of Agriculture), and Harry W. Hetzler (chemist in Youngstown, Ohio). All of the above were Gallaudet College graduates.

3. These people most likely had conductive, rather than sensorineural, hearing losses, in which the sense organ is usually still intact.

4. There are a number of other late-deafened physicians mentioned in the literature, including Cesar Servellon, a native of El Salvador, who lost all of his hearing in the middle of his third year at the University of Wisconsin medical school while preparing for his internship and was turned down for a residency in every hospital he applied to until Prince George's General Hospital viewed him as a top recruit; and Peter Fine who was deafened while a pediatrician practicing in New York in the 1960s. Fine used his experience of deafness and "transformed it into a source of meaning for his life, because he was angry with the way deaf people are treated" (Bernstein, 1975, p. 1). During the late 1980s and early 1990s the Association of Late-Deafened Adults has grown into an effective organization which provides support to people who lose their hearing later in life.

5. Harold Joel Conn Papers (#22/2/661), Department of Manuscripts and University Archives, Cornell University Library.

6. Personal correspondence, Joan Mark, January 3, 1989.

7. Unpublished autobiographical notes sent to me by Olaf Hassel's niece, Mrs. Reidun Guldal, Norway.

Chapter 5

Appreciation is due Earline Sorensen at the National Women's Hall of Fame for her assistance in providing information about Helen Brooke Taussig, and to Vicki Hurwitz for bringing Taussig's deafness to my attention. Thanks also to Donald O. Peterson for sharing his experiences as one of the volunteers in the U.S. Navy experiments on weightlessness; to Simon Carmel for information about Paul S. Watson; and to Judy Tingley for the autobiographical materials of Robert H. Weitbrecht.

1. Several other deaf astronomers should be mentioned. Paul S. Watson attended Baltimore Polytechnic and Johns Hopkins University, from which he graduated in 1929. He held a part-time position at the Maryland Academy of Science observatory, then took a full-time job at the Davis Planetarium, where he lectured to schools and colleges, presented demonstrations, and spent three months each year calculating the academy's annual Graphic Time Table. During his twenty-five years in this position, Watson introduced hundreds of thousands of children, and their parents and teachers, to the field of astronomy.

In Japan, Yukiaki Tanaka, who lost his hearing through meningitis in 1917 at the age of two, attended the Japanese School for Deaf Children and remained there until he was nine. During World War II, he read a newspaper advertisement about a manpower shortage at the Tokyo Observatory and applied for a position. He was assigned the responsibility of observing sunspots and solar prominences with an eight-inch equatorial telescope under Sekiguchi. On the morning of August 15, 1951, Tanaka realized his dream of discovering something for himself. He identified three faculae near the north pole of the sun. Taking an interest in Tanaka's work, Dr. Kuniji Saito introduced the discovery of the polar faculae of the sun to academic circles, and, after being verified, Yukiaki Tanaka's work was eventually included in astronomical texts.

2. Kidd, born deaf, received a bachelor's degree in chemical engineering in 1941, a master's degree in chemistry and geology in 1946, and a Ph.D. from the University of Toronto in 1951. His thesis was titled "The Geochemistry of Beryllium." He joined an expedition to Baffin Island in 1953. The group of thirteen scientists, headed by Colonel Patrick D. Baird, director of the Montreal office of the Arctic Institute, studied geomorphology, botany, and zoology while traveling over the Cumberland Peninsula on foot and on skis. During this ill-fated expedition, W.R.B. Battle from McGill University drowned on July 13 while exploring a glacial stream. Kidd himself met danger several times during the expedition, as Alex Henderson has described: "One day while turning the corner of a glacier moraine, [Kidd] just about walked into the arms of a monster polar bear. They stared at each other for several frightening seconds, then the scientist pursed his quivering lips and shouted 'shoo.' Strangely, the big animal ambled off, but before he disappeared from view the doctor snapped a photo of him" (Undated newspaper clipping, *The Star Weekly*. Gallaudet University Archives Biographical File "Donald J. Kidd."). Contemporary deaf persons with Ph.D.'s in physics include David S. Coco (NASA), Leon N. Kapp (biophysicist at the Laboratory of Radiobiology and Environmental Health at the University of California in San Francisco), and Steven K. Wonnell at the University of North Carolina who, with a Ph.D. in Experimental Condensed Matter Physics, has published studies on point defects in ionic crystals and the Ionic Hall Effect.

3. These data were collected by Professor Francis Higgins while on the chemistry faculty at Gallaudet University and shared with me.

4. One of the first congenitally deaf persons to earn a medical degree outside the United States was Voya Alexander Raykovic, born on February 14, 1924, in Sarajevo, Yugoslavia. Raykovic received his bachelor of arts degree in 1940 from the Classical Gymnasium in Belgrade. He earned the degree of veterinarian in 1950

from the School of Veterinary Medicine in Belgrade, and the magister of bacteriology in 1953 from the Hygiensches Institut der Tierärztlichen Hochschule in Hanover. Raykovic became a citizen of the United States in 1965.

5. Like James, other deaf scientists are currently on the teaching faculty at universities in the United States, including Bruce Hawkins (physics at Smith College in Massachusetts), Fred H. Walters (chemistry, University of Southwestern Louisiana), and Verner C. Johnson (geology, Mesa State College in Colorado). Chuzo Okuda and Angel Ramos are two other minority deaf individuals who have taught mathematics on the postsecondary level. At Gallaudet University, deaf faculty members in biology include Jane Dillehay and Edith Rikuris; in chemistry, Charles A. Giansanti, Michael L. Moore, and Donald O. Peterson; and, in mathematics and computer science, Patrick O. Atuonah, Harvey Goodstein, Patricia Herbold, Vicki J. Kemp, Raymond G. Kolander, Fat C. Lam, Poh-Pin Mangrubang, Herbold G. Mapes, James A. Nickerson, Jr., Marius R. Titus, Marybeth Williamson, and Rudolph C. Hines. Gallaudet's Northwest campus Math Department includes Camilla S. Lange, Jean L. Schickel, and Florence C. Vold. At the National Technical Institute for the Deaf, Dale Rockwell teaches chemisty, Keith Mousley, Robert Menchel, and Sharon Metevier-Webster teach mathematics, Anthony Spiecker teaches in the Electromechanical Technology Program, and Warren Goldmann (mathematics) and Thomas Callaghan (Engineering) are in the Support Departments for cross-registered students.

6. Helen Brooke Taussig was no stranger to attitude barriers. As a child she had to overcome reading difficulties associated with dyslexia. After attending Radcliffe College for two years and earing an A.B. degree from the University of California at Berkeley (1921), she considered Harvard Medical School but encountered opposition from President Lowell, who did not support the admission of women. Taussig completed her M.D. degree at Johns Hopkins Medical School in 1927, but then the department of medicine refused to accept her as an intern. Her subsequent change of interest to pediatric cardiology, however, brought her into collaboration with Alfred Blalock. After investigating birth defects in Germany, Taussig warned the medical community in the United States about the dangers of thalidomide. She received the President's Medal of Freedom in 1964 and was the first woman president of the American Heart Association.

7. The few deaf women engineers that there are include Kathryn Woodcock and Sybella Patten. Woodcock, a systems design engineer from Brantford, Ontario, was honored with the Edmund Lyons Lectureship at the National Technical Institute for the Deaf in 1992. In 1981 she was hired as vice president of hospital services at Centenary Hospital in Scarborough, Ontario, where she was responsible for six departments and staff equivalent to 430 full-time positions. Her current research interests include occupational safety, risk measurement, and accident prevention. Patten is a project engineer at TRW, Electronics and Defense, in Redondo Beach, California.

Among the deaf women scientists listed in the AAAS Directory (Stern, Lifton, & Malcom, 1987) are Sharon L. Campbell, environmental safety and health coordinator; Catherine L. Gatchell, a research chemist in Kalamazoo, Michigan; and Deborah Harrell Saville, a biochemist working as a programmer analyst in Marlboro, Maryland. Deaf women medical doctors include M. Theresa B. San Augustin, Susan Feder, and Karen Pennington (who also holds a Ph.D. in radiobiology). Deaf women scientists who have entered science teaching include Doris Wilson Blanchard, Del Wynne, and Katherine G. Runkle. Sharon S. Chadwick is a science reference librarian with master's degrees in both chemistry and information studies.

8. In addition to the deaf physicians I have thus far mentioned, Myron Weinberger is a professor of medicine and director of the Hypertension Research

Center in Indianapolis, and James McFarland, Jr., is the president of the Western Homeopathic Medical Society in New Jersey. Other dentists include James G. Snyder in Dubuque, Iowa, Bettina Pels-Wetzel in Schenectady, New York, and Christopher Lehfeldt in Rochester, New York.

9. To reinforce the argument that *specialization* within any field makes possible the inclusion of deaf persons, it is interesting to note that at least five deaf persons have entered the field of audiology.

10. A similar cooperative effort, under the coordination of Steven Jamison, at IBM, occurred in the 1980s among IBM, NTID, Gallaudet University, and deaf and hearing computer specialists skilled in sign language. These efforts resulted in the publication of a book that is available from the National Association of the Deaf (*Signs for Computer Terminology*, Steven Jamison, editor) and three companion videotapes available from NAD and Captioned Films for the Deaf. The Technical Signs Project (TSP) was established at the National Technical Institute for the Deaf at Rochester Institute of Technology in 1975 with Dr. Frank Caccamise as TSP Director. Publications include Volume 10 on science signs, with accompanying videotapes.

11. In addition to the scientists I have mentioned, there are approximately 200 deaf men and women listed in the American Association for the Advancement of Science (AAAS) *Resource Directory of Scientists and Engineers with Disabilities*. The project, funded by the National Science Foundation, also includes scientists who are blind and those who have neurological, neuromuscular, organic, orthopedic, or speech disabilities. These scientists are willing to serve as role models, consultants, speakers, or advisors to individuals or groups interested in enhancing educational and employment opportunities for people with disabilities. They are listed by geographic location, disability, scientific specialty, and sex. Approximately 10 percent of the deaf scientists listed are in supervisory positions. This AAAS directory, and biographical data gathered for the present book, provide an effective argument against the belief that deafness introduces insurmountable communication barriers which would preclude supervision. Earlier, I mentioned such deaf scientists as Gideon E. Moore, Robert J. Farquharson, Lucius W. Case, Averill J. Wiley, and others who have supervised staffs of hearing people. In the 1950s and 1960s, other deaf scientists in supervisory positions were reported in the literature of the Deaf community. Raymond T. Atwood, deafened by spinal meningitis at the age of eleven, attended the Louisiana State School for the Deaf and received his bachelor's degree from Gallaudet College and his master's degree from Louisiana State University. As a bacteriologist, he headed research projects on the uses of protein materials left over from brewing operations. He and his staff at the Anheuser-Busch laboratory in St. Louis, Missouri, focused on the production of vitamins and antibiotics. Jean Kelsh Cordano was deafened at the age of four from meningitis. She became the first deaf person to be registered by the American Society of Clinical Pathologists. A graduate of the North Dakota School for the Deaf and Gallaudet College, she completed requirements for the medical technology program at the University of Wisconsin-Madison in 1959 and, after an internship in a hospital, she took a position at the clinical laboratory at Lakeland Hospital in Elkhorn, Wisconsin, where she soon advanced to the position of administrative director.

12. Personal Communication, James C. Marsters, August 11, 1990.

13. For more information about the Gallaudet protest, see Jack R. Gannon (1989), *The week the world heard Gallaudet*. Washington, DC: Gallaudet University Press.

Conclusion

1. In March, 1993, a second working conference was held in Kansas City, Missouri, led by Greg Stefanich, University of Northern Iowa, and George R. Davis, Moorhead State Univeristy, and sponsored by the Science Association for Persons with Disabilities (SAPD), National Science Foundation (NSF), American Association for the Advancement of Science (AAAS), and the Association for the Education of Teachers in Science (AETS). Many of the same attitude problems were found to persist fifteen years after the first conference. At this conference, Harry G. Lang presented a paper titled "Science for Deaf Students: Looking into the Next Millenium" which outlined broad recommendations for science educators of deaf and hard-of-hearing students in both residential and mainstream programs in the United States. Other educators of deaf students who have played key roles in SAPD include Jeff Himmelstein and Judy Egelston-Dodd, both of whom were respondents to Lang's paper.

2. See "A Design for Utilizing Successful Disabled Scientists as Role Models," American Association for the Advancement of Science.

Bibliography

Amman, J. C. (1873) . *A dissertation on speech.* Amsterdam: North Holland.

Anderson, L. (1982). *Charles Bonnet and the order of the known.* Dordrecht, Holland: D. Reidel.

Andrews, H. N. (1980). *The fossil hunters: In search of ancient plants.* Ithaca, NY: Cornell University Press.

Andrews, H. U. (1910). Deaf girls as hospital nurses. *The Volta Review, 12* (8), 471–476.

Appleyard, R. (1931). A link with Oliver Heaviside. *Electrical Communication, 10*, 53–59.

Appleyard, R. (1939). *The history of the Institution of Electrical Engineers (1871–1931).* London: The Institution of Electrical Engineers.

Babbitt, L. H. (1939). The blue-tailed skink in Connecticut. *New England Naturalist, 4,* September.

Bailey, S. I. (1922). Obituary of Henrietta Swan Leavitt. *Popular Astronomy, 30* (4),197–199.

Bailey, S. I. (1931). *History and work of the Harvard Observatory, 1839 to 1927.* New York: McGraw-Hill.

Bates, R. L. (1970). Deaf people in computer professions. *The W. A. D. Pilot,* No. 174, p. 1.

Batson, T., & Bergman, E. (Eds.). (1985). *Angels and outcasts: An anthology of deaf characters in literature.* Washington, DC: Gallaudet College Press.

Belenky, M. F., Clinchy, B. M., Goldberger, N. R., & Tarule, J. M. (1986). *Women's ways of knowing: The development of self, voice, and mind.* New York: Basic Books.

Bell, A. G. (1883). *Memoir upon the formation of a deaf variety of the human race.* New Haven: National Academy of Sciences.

Bell, A. G. (1884). Fallacies concerning the deaf. *American Annals of the Deaf and Dumb, 29* (1), 32–69.

Bell, A. G. (1908). A few thoughts concerning eugenics. *National Geographic Magazine, 19* (2), 119–123.

Bender, R. E. (1981). *The conquest of deafness: A history of the long struggle to make possible normal living to those handicapped by lack of normal hearing*. Danville, IL: Interstate Printers & Publishers.
Benedict, R. (1934). *Patterns of culture*. New York: Houghton Mifflin.
Benedict, R. (1946). *The chrysanthemum and the sword: Patterns of Japanese culture*. Boston: Houghton Mifflin.
Bernstein, B. (1975). Peter Jason Fine, M.D. *The Buff and Blue*, *85* (2), 1.
Bingham, H. B. (1845). *Essays by the pupils at the College of the Deaf and Dumb*. London: Longman.
Boatner, M. T. (1959). *Voice of the deaf: A biography of Edward Miner Gallaudet*. Washington, DC: Public Affairs Press.
Bohart, J. P. (1970). Deaf candidates make strong bid for federal prize. *The Deaf American*, *22* (10), 7–8.
Bonnet, G. (1929). *Charles Bonnet*. Paris.
Bowe, F. (1972). Edwin Nies: Dentist and clergyman, *The Deaf American*, *24* (6), 3–5.
Bowe, F. (1973). Bob Harris interviews Dr. Donald Ballantyne. *The Deaf American*, *26* (2), 5–9.
Braddock, G. C. (1919). Firestone: The new silent colony. *The Silent Worker*, *31* (8), 135–136.
Braddock, G. C. (1975). *Notable deaf persons*. Washington, DC: Gallaudet College Alumni Association.
Brain, W. R. (1964). *Doctors past and present*. Springfield, IL: Charles C. Thomas.
Brasch, F. E. (1931). The Royal Society of London and its influence upon scientific thought in the American colonies. *Scientific Monthly*, *33*, 337–355, 448–469.
Brashear, J. A. (1988). *A man who loved the stars: The autobiography of John A. Brashear*. Pittsburgh: University of Pittsburgh Press.
Brown, R. (1976). Dr. Nansie Sharpless: Biochemist. *The Deaf American*, *28* (7), 3–4.
Bruce, R. V. (1973). *Alexander Graham Bell and the conquest of solitude*. Boston: Little, Brown.
Burkhardt, F., & Smith, S. (Eds.). (1985). *A calendar of the correspondence of Charles Darwin, 1821–1882*. New York: Garland.
Burlingame, R. (1964). *Out of silence into sound: The life of Alexander Graham Bell*. New York: Macmillan.
Burnham, C. (1924). Deaf and dumb, Milwaukee doctor overcomes handicap with optimism. *The Silent Worker*, *37* (3), 99–100.
Caffrey, M. M. (1989). *Ruth Benedict: Stranger in this land*. Austin: University of Texas Press.
Cajori, F. (1962). *A history of physics in its elementary branches (through 1925): Including the evolution of physical laboratories*. New York: Dover.
Campbell, L. (1941). Annie Jump Cannon. *Popular Astronomy*, *49* (7), 345–347.

Carmel, S. (1987). Folklore. In J. V. Van Cleve (Ed.), *Gallaudet encyclopedia of deaf people and deafness* (Vol. 1, pp. 428–430). New York: McGraw-Hill.

Chace, F. A. (1947). Obituary of Hubert Lyman Clark. *Science, 106* (2740), 611–612.

Christiansen, J. (1987). Minorities. In J. V. Van Cleve (Ed.), *Gallaudet encyclopedia of deaf people and deafness* (Vol. 1, pp. 270–276). New York: McGraw-Hill.

Clendenning, L. (1933). *Behind the doctor.* London: William Heinemann.

Corbett, E. E., & Jensema, C. J. (1981). *Teachers of the deaf: Descriptive profiles.* Washington, DC: Gallaudet College Press.

Cornforth, J. W. (1976). Asymmetry and enzyme action. *Science, 193* (4248), 121–125.

Costain, T. B. (1960). *The chord of steel: The story of the invention of the telephone.* Garden City, NY: Doubleday.

Crammatte, A. B. (1968). *Deaf persons in professional employment.* Springfield, IL: Thomas.

Crammatte, A. B. (1987). Wilson H. Grabill. In J. V. Van Cleve (Ed.), *Gallaudet encyclopedia of deaf people and deafness* (Vol. 1, pp. 477–478). New York: McGraw-Hill.

Crammatte, F. (1987). Regina Olson Hughes. In J. V. Van Cleve (Ed.), *Gallaudet encyclopedia of deaf people and deafness* (Vol. 2, pp. 77–78). New York: McGraw-Hill.

Crowe, M. J. (1967). *A history of vector analysis: The evolution of the idea of a vectorial system.* Notre Dame, IN: University of Notre Dame Press.

Crowther, J. G. (1937). *Famous American men of science.* New York: W. W. Norton.

Crowther, J. G. (1970). *Fifty years with science.* London: Barrie & Jenkins.

Currier, E. H. (1912). *The deaf: By their fruits ye shall know them.* New York: New York Institution for the Instruction of the Deaf and Dumb.

D'Ans, J. (1931). Carl Freiherr Auer von Welsbach. *Berichte der Deutschen Chemischen Gesellschaft, 64* (6), 59–92.

Darrah, W. C. (1934). Leo Lesquereux. *Harvard University Botanical Museum Leaflets, 2* (10), 113–119.

Darwin, F. (1887). *Life and letters of Charles Darwin.* London.

Darwin, F., & Seward, A. C. (Eds.). (1903). *More letters of Charles Darwin: A record of his work in a series of hitherto unpublished letters.* New York: D. Appleton.

Davenport, C. B. (1939). Frederick Augustus Porter Barnard. *National Academy of Sciences Biographical Memoirs* (Vol. 20, pp. 259–272). Washington, DC: National Academy of Sciences.

Davies, L. A. (1924). Lucius W. Case—bacteriologist. *The Volta Review, 26* (3), 122–127.

Davis, H., & Silverman, S. R. (1978). *Hearing and deafness.* New York: Holt, Rinehart and Winston.

de Caraman, Le Duc. (1859). *Charles Bonnet, philosophe et naturaliste: Sa vie et ses oeuvres.* Paris: A. Vaton, Libraire-Éditeur.

DeLand, F. (1924). Who was the tenth century teacher, "John the Deaf"? *The Volta Review, 26* (10), 535.

Dewhurst, K. (1963). *John Locke (1632–1704): Physician and philosopher.* London: The Wellcome Historical Medical Library.

Dostrovsky, S. (1975). Joseph Sauveur. In C. C. Gillispie (Ed.), *Dictionary of scientific biography* (Vol. 12, pp. 127–129). New York: Charles Scribner's Sons.

DuBow, S. (1987). Davis v. Southeastern Community College. In J. V. Van Cleve (Ed.). *Gallaudet encyclopedia of deaf people and deafness* (Vol. 2, pp. 412–414). New York: McGraw-Hill.

Dunlap, O. E. (1944). *Radio's 100 men of science.* New York: Harper & Brothers.

Dunlop, S. (1981). John Goodricke: 1764–1786. *Soundbarrier, 9* (June), 13–14.

Dunn, V. (1924). In memoriam. *The Silent Worker, 36* (8), 354–357.

Durand, W. F. (1938). Harris Joseph Ryan. *National Academy of Sciences Biographical Memoirs* (Vol. 18, pp. 285-306). Washington, DC: National Academy of Sciences.

Duster, T. (1990). *Backdoor to eugenics.* New York: Routledge.

Eagar, R. M. (1947, May). I made a career: Geologist, curator and lecturer. *The Silent World*, pp. 326–327, 349.

Eccles, W. H. (1945). Obituary of John Ambrose Fleming. *Nature, 155* (3944), 662–663.

Eddy, J. A. (1972). Thomas A. Edison and infra-red astronomy. *Journal for the History of Astronomy, 3*, 165–187.

Elliott, C. A. (1990). Collective lives of American scientists: An introductory essay and a bibliography. In E. Garber (Ed.), *Beyond history of science: Essays in honor of Robert E. Schofield.* Cranbury, NJ: Associated University Presses.

Elrel, E. L., & Mosher, H. S. (1975). The 1975 Nobel Prize for chemistry. *Science, 190* (4216), 772–774.

Farquhar, G. C. (1920). The deaf in industrial rubber chemistry. *The Silent Worker, 32* (6).

Farrar, A. (1939, July). My story. *Digest of the Deaf.*

Fay, E. A. (1898). *Marriages of the deaf in America: An inquiry concerning the results of marriages of the deaf in America.* Washington, DC: The Volta Bureau.

Fechter, A. (1989). A statistical portrait of black Ph.D.s. In W. Pearson, Jr., & H. K. Bechtel (Eds.), *Blacks, science, and American education.* New Brunswick, NJ: Rutgers University Press.

Ferris, T. (1979). Voyager's music. In C. Sagan, F. D. Drake, A. Druyan, T. Ferris, J. Lomberg, & L. Salzman Sagan, *Murmurs of Earth: The Voyager interstellar record.* New York: Ballantine Books.

Fischer, L. J., & de Lorenzo, D. L. (Eds.). (1983). *History of the college for the deaf: 1857–1907 by Edward Miner Gallaudet.* Washington, DC: Gallaudet College Press.

Fleming, D. (1971). Ruth Benedict. In *Notable American women, 1607–1950: A biographical dictionary* (Vol. 1, pp. 128–131). Cambridge: Harvard University Press.

Fleming, J. A. (1934). *Memories of a scientific life.* London: Marshall, Morgan & Scott.

Fontenelle, Bernard Le Bovier de. (1705). Éloge de M. Amontons. In *Histoire de l'Académie Royale des Sciences* (pp. 150–154).

Ford, H. (1930). *Edison as I know him.* In collaboration with S. Crowther. New York: Cosmopolitan Book.

Freire, P. (1970). *Pedagogy of the oppressed.* New York: Herder & Herder.

Friedel, R. (1986). *Edison's electric light: Biography of an invention.* New Brunswick, NJ: Rutgers University Press.

Fulton, J. (1896). *Memoirs of Frederick A. P. Barnard, D.D., L.L.D., L.H.D., D.C.L., Etc., tenth president of Columbia College in the city of New York.* New York: Macmillan.

Gacs, U., Kahn, A., McIntyre, J., & Weinberg, R. (Eds.). (1989). *Women anthropologists: Selected biographies.* Chicago: University of Illinois Press.

Gallaher, J. E. (1898). *Representative deaf persons of the United States of America.* Chicago: Gallaher.

Gallaudet, E. M. (1893). *The Columbia Institution for the Instruction of the Deaf and Dumb: 1857–1893.* Washington, DC: The National Deaf-Mute College.

Gannon, J. R. (Ed.). (1981). *Deaf heritage: A narrative history of deaf America.* Silver Spring, MD: National Association of the Deaf.

Garretson, M. D. (1950). The versatile . . . Edwin W. Nies. *The Silent Worker, 3* (2), 3–5.

Gaston, J. (1989). The benefits of black participation in science. In W. Pearson, Jr., & H. K. Bechtel (Eds.), *Blacks, science, and American education* (pp. 123–136). New Brunswick, NJ: Rutgers University Press.

Gavin, J. (1975). Who disables the disabled? *Chemical Technology, 5* (December), 716–721.

Gavin, J. (1979a, March). Some random thoughts on the education of the handicapped with particular reference to the deaf. *Foundation for Science and the Handicapped Newsletter.*

Gavin, J. (1979b). Judicial reasoning or attitudinal barriers? *The Deaf American, 32* (1), 7–9.

Gilman, C. (1978). John Goodricke and his variable stars. *Sky and Telescope, 56* (3), 400–403.

Gingerich, O. (1964). Laboratory exercises in astronomy—spectral classification. *Sky and Telescope, 28* (August), 80–82.

Golladay, L. E. (1962). Goodricke story includes locating of his observatory; memorial fund will honor him. *The American Era, 48* (4), 33–37.

Goodricke, J. (1783). A series of observations on, and a discovery of, the period of variation of light of the bright star in the head of Medusa, called Algol. *Philosophical Transactions of the Royal Society, 73*, 474–482.

Goodricke, J. (1785). Observations of a new variable star. *Philosophical Transactions of the Royal Society, 74*, 287–292.

Gotthelf, J. F. (1919). The value and efficiency of deaf labor to the industrial world. *The Buff and Blue, 28* (1), 9–11.

Gould, S. J. (1980). Tilly Edinger. In B. Sicherman, C. H. Green, I. Kantrov, & H. Walker (Eds.), *Notable American women: The modern period* (Vol. 4, pp. 218–219). Cambridge: Belknap Press of Harvard University Press.

Gould, S. J. (1988). Pretty pebbles. *Natural History, 97* (4), 14,16.

Graybill, G. & Petersen, E. (1963). The silent angels of Mercy Hospital. *The Silent Worker, 15* (5), 28–29.

Green, F. (1783). *Vox oculis subjecta.* London: Benjamin White.

Grigorian, A. T. (1976). Konstantin Eduardovich Tsiolkovsky. In C. C. Gillispie (Ed.), *Dictionary of scientific biography* (Vol. 13, pp. 482–484). New York: Charles Scribner's Sons.

Grmek, M. D. (1978). Charles Jules Henri Nicolle. In C. C. Gillispie (Ed.), *Dictionary of scientific biography* (Vol. 15, pp. 453–455).

Groves, E. W. (1976). Gilbert White. In C. C. Gillispie (Ed.), *Dictionary of scientific biography* (Vol. 14, pp. 299–300). New York: Charles Scribner's Sons.

Hairston, E., & Smith, L. (1983). *Black and deaf in America: Are we that different.* Silver Spring, MD: T.J. Publishers.

Halle, T. G. (1921). Alfred Gabriel Nathorst. *Geologiska föreningens i Stockholm förhandligar, 43,* 241–280.

Harshberger, J. W. (1899). *The botanists of Philadelphia and their work.* Philadelphia: T. C. Davis & Sons.

Hill, W. (1895). Glimpses of the past: Reminiscences of the sixties. *The Buff and Blue, 3,* (April 12), 5.

Hoag, R. L. (1989). *The origin and establishment of the National Technical Institute for the Deaf.* Rochester, NY: National Technical Institute for the Deaf.

Hochman, F. P. (1986). You know I can't hear you when the I. V. is running. Edmund Lyon Memorial Lectureship. National Technical Institute for the Deaf at Rochester Institute of Technology.

Hodgson, K. W. (1953). *The deaf and their problems.* New York: Philosophical Library.

Hofer, H. (1969) In memoriam Tilly Edinger. *Gegenbaurs Morphologisches Jahrbuch, 113* (2), 303–313.

Hoffleit, D. (1971a). Annie Jump Cannon. In E. T. James, J. W. James, & P. S. Boyer (Eds.), *Notable American Women* (Vol. 1). Cambridge: Harvard University Press.

Hoffleit, D. (1971b). Henrietta Swan Leavitt. In E. T. James, J. W. James, & P. S. Boyer (Eds.), *Notable American Women* (Vol. 2). Cambridge: Harvard University Press.

Holcomb, M., & Wood, S. K. (1989). *Deaf women: A parade through the decades.* Berkeley, CA: Dawn Signs Press.

Holt, L. D. (1925). A dreamer of practical dreams. *The Volta Review, 27* (12), 675–682.

Holt, L. D. (1927). An Edison of the Pacific coast. *The Volta Review, 29* (5), 209–213.

Holt-White, R. (1901). *The life and letters of Gilbert White of Selborne.* London: John Murray.

Horgen, R. W. (1963). Deaf man heads large Wisconsin laboratory. *The Silent Worker, 15* (8), 3–5.

Hoskin, M. (1982). Goodricke, Pigott and the quest for variable stars. In M. Hoskin (Ed.), *Stellar Astronomy: Historical studies.* Chalfont, St. Giles, Bucks, England: Science History Publications.

Hotchkiss, D. (1989). *Demographic aspects of hearing impairments: Questions and answers.* Washington, DC: Gallaudet University.

Jacobs, L. M. (1974). *A deaf adult speaks out.* Washington, D.C.: Gallaudet College Press.

Johnson, S. (1990). *A journey to the western islands of Scotland.* London: The Folio Society.

Johnson, W. (1928). *Gilbert White: Pioneer, poet and stylist.* London: John Murray.

Jones, B. Z., & Boyd, L. G. (1971). *The Harvard college observatory: The first four directorships, 1839–1919.* Cambridge: The Belknap Press of the Harvard University Press.

Kelly-Jones, N. (1974). Where are our deaf women? *Gallaudet Today, 4* (3), 24–29.

Kenschaft, P. C. (1987). Charlotte Angas Scott. In L. S. Grinstein & P. J. Campbell (Eds.),*Women of mathematics: A biobibliographic sourcebook.* New York: Greenwood Press.

Konglia Vetenskaps Academiens Handlingar. (1813). Biography of Magister Anders Gustaf Ekeberg, teacher of the academy and laboratory chemist in Uppsala.

Kopal, Z. (1986). *Of stars and men: Reminiscences of an astronomer.* Boston: Adam Hilger.

Kosmodemyansky, A. (1956). *Konstantin Tsiolkovsky: His life and work.* Moscow: Foreign Languages Publishing House.

Kroeber, T. (1970). *Alfred Kroeber: A personal configuration.* Berkeley: University of California Press.

Lane, H. (1984). *When the mind hears: A history of the deaf.* New York: Random House.

Lang, H. G., & Propp, G. (1982). Science education for hearing-impaired students: State of the art. *American Annals of the Deaf, 127* (7), 860–869.

Lang, H. G., & Meath-Lang, B. (1985, June). Hearing-impaired students' attitudes toward science: Implications for teachers. Paper presented at the Convention of American Instructors of the Deaf, St. Augustine, FL.

Langrall, P. (1987, January). Regina Hughes lends her skills to SI botanists. Smithsonian Institution *Torch*, p. 5.

Lash, J. P. (1980). *Helen and teacher: The story of Helen Keller and Anne Sullivan Macy.* New York: Delacorte Press.

Lesley, J. P. (1890). Memoir of Leo Lesquereux. *National Academy of Sciences Biographical Memoirs* [Vol. 3, pp. 189-212). Washington, DC: National Academy of Sciences.

Levine, R. (1991). Deaf survivors' stories told at last. *Silent News, 23* (1), 27.

Ley, W. (1966). *Watchers of the skies: An informal history of astronomy from Babylon to the space age*. New York: The Viking Press.

Lo, A. H. (1978). An interview with Professor John W. Cornforth. *The Deaf American, 30* (6), 3–8.

Logan, J. H. (1877). The necessity of a training-school for teachers of deaf-mutes. *American Annals of the Deaf and Dumb, 22* (2), 89–92.

Lot, G. (1961). *Charles Nicolle et la biologie conquérante*. Paris.

Lubbock, C. A. (Ed.). (1933). *The Herschel chronicle: The life-story of William Herschel and his sister Caroline Herschel*. Cambridge, England: The University Press.

Lubbock, J. (1872). *Monograph of the Collembola and Thysanura*. London: J. E. Adlard.

Mabee, C. (1969). *The American Leonardo: A life of Samuel F. B. Morse*. New York: Octagon Books.

MacGregor-Morris, J. T. (1954). *The inventor of the valve: A biography of Sir Ambrose Fleming*. London: The Television Society.

MacGregor-Morris, J. T. (1959). John Ambrose Fleming. In L. G. Wickham Legg & E. T. Williams (Eds.), *The dictionary of national biography 1941–1950* (pp. 258–260). London: Oxford University Press.

MacLeod-Gallinger, J. E. (1992). The career status of deaf women. *American Annals of the Deaf, 137* (4), 315–325.

Maestas y Moores, J., & Moores, D. F. (1984). The status of Hispanics in special education. In G. L. Delgado (Ed.), *The Hispanic deaf: Issues and challenges for bilingual special education*. Washington, DC: Gallaudet College Press.

Mandula, B. (1987). Science teachers at Gallaudet University. *Association for Women in Science Newsletter, 16* (4), 6–9.

Marnette, M. (1927). A successful chemist. *The Volta Review, 29* (3), 70–72.

Martin, M. (1899, July 11-14). Employments open to deaf women. *Proceedings of the Sixth Convention of the National Association of the Deaf*, St. Paul, MN, 76–81.

McCabe, L. R. (1887). Sketch of Leo Lesquereux. *Popular Science Monthly, 30*, 835–840.

McClure, G. M., (1962). Death takes George Bryan Shanklin, Deaf General Electric cable expert. *The Silent Worker, 14* (6), 7.

McGourty, F., Jr. (1967). Thomas Meehan, 19th century plantsman. *Plants and Gardens, 23* (Autumn), 81, 85.

McLain, S. J., & Perkins, C. O. (1990, March). Disabled women: At the bottom of the work heap. *Vocational Educational Journal*, pp. 53–54.

Mead, M. (1966). *An anthropologist at work: Writings of Ruth Benedict*. Boston: Houghton Mifflin.

Mead, M. (1974). *Ruth Benedict: A humanist in anthropology*. New York: Columbia University Press.

Meehan, T. (1883). Variations in nature: A contribution to the doctrine of evolution, and the theory of natural selection. *Proceedings of the American Association for the Advancement of Science Thirty-first meeting held at Montreal, Canada, August, 1882.* Salem: Published by the Permanent Secretary.

Meeron, E. (1951). Emmanuel Meeron: An autobiographical sketch. *The Volta Review, 53* (10), 457–458, 490.

Melmore, S. (1949). The site of John Goodricke's observatory. *Observatory, 69* (850), 95–99.

Menchel, R. S., & Ritter, A. (1984, August). Keep deaf workers safe. *Personnel Journal*, pp. 49–51.

Merrill, G. P. (1924). *The first one hundred years of American geology.* New Haven: Yale University Press.

Mettler, C. C. (1947). *History of medicine.* Philadelphia: Blakiston.

Miall, L. C. (1912). *The early naturalists: Their lives and work (1530–1789).* New York: Macmillan.

Middleton, W. D. (1885). Biographical sketch of Dr. Robert James Farquharson. *Proceedings of the Davenport Academy of Sciences.* Davenport, IA.

Middleton, W. E. K. (1968). *The history of the barometer.* Baltimore: Johns Hopkins University Press.

Milner, L. B. (1965). Dr. Donald Ballantyne—deaf researcher in transplants. *The Deaf American, 18* (3), 3–4.

Modell, J. S. (1983). *Ruth Benedict: Patterns of a life.* Philadelphia: University of Pennsylvania Press.

Moore, D. H. (1971). *Heaviside operational calculus: An elementary foundation.* New York: American Elsevier.

Moore, G. E. (1865). On Brushite, a new mineral occurring in phosphatic guano. *American Journal of Science, 39*, 44.

Moore, N. T. (1987, Spring). On their way up—thanks in part to the GFF. *Gallaudet Today.*

Moores, D. F. (1987). Higher education. In J. V. Van Cleve (Ed.), *Gallaudet encyclopedia of deaf people and deafness* (Vol. 1, pp. 403–404). New York: McGraw-Hill.

Nahin, P. J. (1988). *Oliver Heaviside, sage in solitude: The life, work, and times of an electrical genius of the Victorian age.* New York: Institute of Electrical and Electronics Engineering.

National Information Center on Deafness. (1987). *Career information registry of hearing impaired persons in professional, technical, and managerial occupations.* Washington, DC: Gallaudet University.

National Science Foundation. (1990). *Report of the National Science Foundation Task Force on Persons with Disabilities.* Washington, DC: National Science Foundation.

Newcomb, S. (1903). *Reminiscences of an astronomer.* Boston: Houghton Mifflin.

Nitchie, E. B. (1916). A deaf man's music. *The Volta Review, 18* (9), 373–374.

Nobel Foundation. (1975). *Le Prix Nobel.* Stockholm: Imprimerie Royale.

Nordenskiöld, E. (1935). *The history of biology: A survey.* New York: Tudor.

Oehser, P. H. (1968). *Sons of science: The story of the Smithsonian Institution and its leaders.* New York: Greenwood Press.

Ogilvie, M. B. (1986). *Women in science: Antiquity through the nineteenth century.* Cambridge: MIT Press.

Orton, E. (1890). Leo Lesquereux. *American Geologist, 5,* 284–296.

Padden, C. & Humphries, T. (1988). *Deaf in America: Voices from a culture.* Cambridge: Harvard University Press.

Page, T., & Page, L. W. (Eds.). (1976). *Space science and astronomy: Escape from earth.* New York: Macmillan.

Palmer, A. J. (1984). *Edison: Inspiration to youth.* Milan, OH: Edison Birthplace Association.

Panara, J. (1987). George E. Hyde. In J. V. Van Cleve (Ed.), *Gallaudet encyclopedia of deaf people and deafness* (Vol. 2, pp. 79–80). New York: McGraw-Hill.

Panara, R. (1954). The deaf writer in America: 1900–1954. *The Silent Worker, 7* (2), 3–5.

Panara, R. (1970). The deaf writer in America from colonial times to 1970. Part I. *American Annals of the Deaf, 115* (5), 509–513.

Panara, R. (1974). The deaf writer in America from colonial times to 1970. Part II. *American Annals of the Deaf, 115* (7), 673–679.

Panara, R., & Panara, J. (1983). *Great deaf Americans.* Silver Spring, MD: T. J. Publishers.

Payne-Gaposchkin, C. H. (1941, May–June). Miss Cannon and stellar spectroscopy. *The Telescope,* pp. 62–63.

Pearl, R. (1927, November). The biology of superiority. *American Mercury, 12.*

Pearson, W. (1985). *Black scientists, white society, and colorless science: A study of universalism in American science.* Millwood, NY: Associated Faculty Press.

Pearson, W., Jr., & Bechtel, H. K. (1989). (Eds.). *Blacks, science, and American education.* New Brunswick, NJ: Rutgers University Press.

Petersen, E. W. (1974). Wenger twins honored with Gallaudet doctorates. *The Deaf American, 26* (5), 3–4.

Petersen, E. W. (1975). Hoosier Bob Bates—outstanding mathematician, programmer. *The Deaf American, 27* (11), 3–6.

Peterson, P. N. (1930, May). A dream—and a possibility. *The Vocational Teacher, 1,* p. 1.

Porter, C. S. (1896). Ormond E. Lewis: A deaf civil engineer. *The British Deaf-Mute, 5* (52), 105.

Pratt, W. S. (1935). *The history of music: A handbook and guide for students.* New York: G. Schirmer.

Rae, L. (1852). On the proper use of signs in the instruction of the deaf and dumb. *American Annals of the Deaf and Dumb, 5* (1), 21–31.

Raman, V. V. (1973). Sauveur, the forgotten founder of acoustics. *The Physics Teacher, 11* (3), 161–163.

Redden, M. R., Davis, C. A., & Brown, J. W. (1978). *Science for handicapped students in higher education.* Washington, DC: American Association for the Advancement of Science.

Riabchikov, E. (1971). *Russians in space.* Garden City, NY: Doubleday.

Richardson, R. S. (1967). *The star lovers.* NewYork: Macmillan.

Robinson, R. (1950). Address at the Heaviside Centenary Meeting. *The Heaviside Centenary Volume.* London: The Institute of Electrical Engineers.

Rodgers, A. D. (1944). *American botany, 1873–1892: Decades of transition.* Princeton, NJ: Princeton University Press.

Roe, W. R. (1917). *Peeps into the deaf world.* Derby: Bemrose.

Rogerson, E. (1949). The Wenger twins today. *The Silent Worker, 1* (6), 3–4.

Romero, E. (1939). The wizard of Menlo Park (Thomas A. Edison). *The Modern Silents, 3* (2), 7.

Rosser, S. V. (1992, September/October). The gender equation. *The Sciences,* pp. 42–47.

Rossiter, M. W. (1982). *Women scientists in America: Struggles and strategies to 1940.* Baltimore: Johns Hopkins University Press.

Rostand, J. (1966). *Hommes d'autrefois et d'aujourd'hui: Charles Bonnet.* Paris.

Round, B. F. (1929). A most extraordinary deaf inventor. *The Iowa Hawkeye, 49* (12), 1.

Runes, D. D. (1948). (Ed.). *The diary and sundry observations of Thomas Alva Edison.* New York: Philosophical Library.

Russell, A. (1925). Obituary of Oilver Heaviside. *Nature, 115* (2885), 237–238.

Sarton, G. (1942). Lesquereux (1806–89). *Isis, 34,* 97–108.

Savioz, R. (1948). *Mémoires autobiographiques de Charles Bonnet de Genève.* Paris.

Schein, J. D. (1975). The deaf scientist. *Journal of Rehabilitation of the Deaf, 9* (1), 17–21.

Schorsch, E. (1934). Eugenics and deaf mutism. *The Volta Review, 36* (6), 329, 375–376.

Schowe, B. M., Jr. (1979). *Identity crisis in deafness.* Tempe, AZ: The Scholars Press.

Schulte, K., & Cremer, I. (1991). Developing structures to serve hearing-impaired students in higher education. Heidelberg: Research Center of Applied Linguistics for the Rehabilitation of the Disabled, Pädagogische Hochschule Heidelberg. Translated by S. Blod.

Searle, G. F. C. (1950). Oliver Heaviside: A personal sketch. *The Heaviside centenary volume.* London: The Institution of Electrical Engineers.

Sharpless, N. S. (1975, April 29). The deaf scientist as a researcher. Paper presented to the American Society for Microbiology, New York.

Sheridan, L. C. (1875). The higher education of deaf-mute women. *American Annals of the Deaf and Dumb, 20* (4), 248–252.

Smaltz, W. M. (1949). An outline of a varied life. *The Silent Worker, 1* (11), 7–8.

Smith, A. M. (1909). Leo Lesquereux: 1806–1889. *The Bryologist, 5* (September), 75–78.

Smith, J. M. (1967). Dr. Anthony A. Hajna: Hoosier bacteriologist. *The Deaf American, 19* (10), 3–6.

Smith, R. A. (1914). The veil of silence. *The Volta Review, 16* (4), 198–207.

Stansfield, R. E. (1947). Hunting amphibia and reptiles. *The American Era, 34* (1).

Stern, V. W., Lifton, D. E., & Malcom, S. M. (1987). *Resource directory of scientists and engineers with disabilities.* Washington, DC: American Association for the Advancement of Science.

Sternberg, C. H. (1931). *The life of a fossil hunter.* San Diego: Jensen Printing.

Swift, J. (1959). *Gulliver's Travels.* New York: E. P. Dutton.

Tanaka, Y. (1976, Summer). The spiritual autobiography of a Christian astronomer. *The Japan Christian Quarterly*, pp. 169–172.

Thayer, W. H. (1862). *Working and winning or, the deaf boy's triumph.* Boston: Henry Hoyt.

Timbs, J. (1860). *Stories of inventors and discoverers in science and the useful arts.* London: Kent.

Tobien, H. (1968). Tilly Edinger. *Paläontologische Zeitschrift, 42*, 1–2.

True, F. W. (Ed.). (1913). *A history of the first half-century of the National Academy of Sciences: 1863–1912.* Washington, DC: National Academy of Sciences.

Tsiolkovsky, K. (1960). *Beyond the planet Earth.* (K. Syers, Trans.). New York: Pergamon Press.

Tully, N. L., & Vernon, M. (1965). The impact of automation on the deaf worker. *The Deaf American, 17* (11), 3–4.

Uglow, J. S. (Ed.). (1982). *The international dictionary of women's biography.* New York: Macmillan.

U.S. Department of Education (1990). *Twelfth annual report to Congress on the implementation of the Handicapped Act.* Division of Innovation and Development, Office of Special Education Programs.

Van Cleve, J. V. (1987). (Ed.). *Gallaudet encyclopedia of deaf people and deafness.* New York: McGraw-Hill.

Van Cleve, J. V., & Crouch, B. A. (1989). *A place of their own: Creating the deaf community in America.* Washington, DC: Gallaudet University Press.

Van Den Bos, W. H. (1952). *Monthly Notices of the Royal Astronomical Society, 112* (3), 271–273.

Van Den Bos, W. H. (1958). Robert Grant Aitken. In *National Academy of Sciences Biographical Memoirs* (Vol. 32). Washington, DC: National Academy of Sciences.

Vogt, G. (1939). Jean Jacques Rousseau—passionate pilgram. *The Volta Review, 41*, 29–30, 60.

von Siemens, W. (1893). *Personal recollections of Werner von Siemens.* New York: D. Appleton.

von Welsbach, Auer (1902, May 30). History of the invention of incandescent gas-lighting. *Chemical News*, pp. 254–256.

Vorobyov, B. (1940). *Tsiolkovsky.* Moscow: Young Guard Publishing House.

Wagner, W. (1976). *Reuben Fleet and the story of Consolidated Aircraft.* Fallbrook, CA: Aero Publishers..

Waller, A. E. (1945). Leo Lesquereux—Swiss American. *The Volta Review, 47* (4), 217–219, 254.

Wandrey, C. (1942). *Werner Siemens: Geschicte seines lebens und wirkens.* München: Albert Langen-George Müller.

Waring, E. S. (1896). *The deaf in the past and present times.* Grinnell, IA: E. S. Waring.

Waterfield, R. L. (1941). Obituary of Annie Jump Cannon. *Nature, 147* (June 14), 738.

Weeks, M. E. (1956). *Discovery of the elements.* Easton, PA: Journal of Chemical Education.

West, A. F. (1892). *Alcuin and the rise of the Christian schools.* Reproduced in 1969 by University Microfilms, Inc., Ann Arbor, MI.

White, C. A. (1896). Biographical sketch of Fielding Bradford Meek. *The American Geologist, 18* (6), 337–350.

Wilczek, F., & Devine, B. (1987). *Longing for the harmonies: Themes and variations from modern physics.* New York: W. W. Norton.

Williams, T. I. (Ed.). (1982). *A biographical dictionary of scientists.* New York: John Wiley and Sons.

Willis, S. E. (1959). Trials and tribulations of a deaf engineer. *The Silent Worker, 11* (12), 3–5.

Winefield, R. (1987). Alexander Graham Bell. In J. V. Van Cleve (Ed.), *Gallaudet encyclopedia of deaf people and deafness* (Vol. 1, pp. 135–141). New York: McGraw-Hill.

Witt, J., & Ogden, P. W. (1981). Politics and deaf people: Part I. *The Deaf American, 33* (10), 5–10.

Wright, D. (1969). *Deafness.* New York: Stein and Day.

Yost, E. (1943). Annie Jump Cannon. *American women of science.* Philadelphia: Frederick A. Stokes.

Young, J. L. (1890). *Edison and his phonograph.* Bournemouth: The Talking Machine Review.

Index

About the Author

HARRY G. LANG is Professor of Educational Research and Development at the National Technical Institute for the Deaf, Rochester Institute of Technology. He has written more than 35 articles and essays on science education and Deaf studies.

ISBN 0-89789-368-9
9 780897 893688
90000
EAN
HARDCOVER BAR CODE